Advanced Methods in

CELLULAR IMMUNOLOGY

Rafael Fernandez-Botran, Ph.D.
Václav Větvička, Ph.D.

CRC Press
Boca Raton London New York Washington, D.C.

Library of Congress Cataloging-in-Publication Data

Catalog record is available from the Library of Congress.

Direct all inquiries to CRC Press LLC, 2000 N.W. Corporate Blvd., Boca Raton, Florida 33431.

For Product Safety Concerns and Information please contact our EU representative:
GPSR@taylorandfrancis.com
Taylor & Francis Verlag GmbH, Kaufingerstraße 24, 80331 München, Germany.

ISBN-13: 9780849321252 (pbk)
ISBN-13: 9781138402287 (hbk)

Preface

Immunologists, as well as investigators in other disciplines, often need to use protocols involving the isolation, culture, or characterization of different types of leukocytes; they may need to study cytokine expression by different cells or use a variety of animal models of inflammation and/or autoimmunity. With the same idea as that of our previous book, *Methods in Cellular Immunology*, we have compiled in this book techniques in the above areas and presented them in an easy-to-use format. Each chapter provides readers with related background information and a step-by-step description of the methodology, alternative techniques, pertinent references, and information about commercial sources for materials and reagents.

This book focuses on the isolation and functional characterization of different types of leukocytes, cell culture, and different animal models of inflammation and autoimmunity. Chapter 1 covers the preparation of cell suspensions from several sources, including bone marrow, thymus, and secondary lymphoid organs, together with culture techniques and protocols to measure cell proliferation. Chapters 2 and 3 are dedicated to the isolation and functional assessment of neutrophils and eosinophils, respectively, including techniques based on density-gradient centrifugation and the more novel immunomagnetic bead isolation protocols. Chapter 4 describes the isolation of human basophils. Chapter 5 examines the isolation and functional characterization of monocytic cells, including assays for the production of oxidative metabolism intermediates, phagocytosis, pinocytosis and bactericidal/antiviral activities. Chapter 6 focuses on B lymphocyte cloning and culture methods, and Chapter 7 explains general techniques in cell culture. Chapter 8 describes the techniques of reverse transcriptase polymerase chain reaction (RT-PCR) and ribonuclease protection assay (RPA), and their applications in the study of cytokine mRNA expression. Chapters 9 and 10 describe different techniques for the intracellular detection of cytokines and the measurement of cell cytotoxicity and apoptosis. Chapter 11 explains mixed lymphocyte reaction. Chapter 12 outlines animal models of autoimmune diseases. Chapter 13 explains the preparation of several common reagents. Chapter 14 provides a list of manufacturers/commercial sources, including addresses, telephone numbers, and Web sites.

The Authors

Rafael Fernandez-Botran, Ph.D., is an associate professor in the Department of Pathology and Laboratory Medicine, Division of Experimental Immunology and Immunopathology, School of Medicine, University of Louisville, Kentucky. Dr. Fernandez-Botran received his bachelor of science degree in biological chemistry from San Carlos University, Guatemala in 1979 and his Ph.D. in microbiology from the University of Kansas Medical Center, Kansas City in 1985. Dr. Fernandez-Botran worked as a postdoctoral fellow and instructor in the laboratories of Dr. Ellen S. Vitetta and Dr. Jonathan W. Uhr at the Department of Microbiology, University of Texas Southwestern Medical Center, Dallas, where his work centered on the study of cytokines, particularly interleukin-4, and their roles in the proliferation and differentiation of $CD4^+$ T cell subsets and B cells, and on the characterization of interleukin-4 receptors and their interactions with interleukin-2 receptors.

More recently, Dr. Fernandez-Botran's major research interests have focused on the immunoregulatory roles of soluble cytokine receptors and the interactions between cytokines and glycosaminoglycans. Dr. Fernandez-Botran has received several fellowships and awards, including the New Investigator Recognition Award from the Clinical Immunology Society (CIS) in 1989, and the President's Young Investigator Award from the University of Louisville in 1993. His research activities have resulted in more than 50 published works, including articles, several reviews, and chapters.

Václav Větvička, Ph.D., is an assistant professor in the Department of Pathology, Division of Experimental Immunology and Immunopathology, School of Medicine, University of Louisville, Kentucky. Dr. Větvička graduated in 1978 from Charles University in Prague, Czech Republic with an advanced degree in biology. He obtained his Ph.D. in 1983 from the Czechoslovak Academy of Sciences, Institute of Microbiology, Prague. He is a member of the Czech Immunological Society, American Association of Immunologists, International Society of Developmental and Comparative Immunology, and American Association for Cancer Research. In 1984 he was awarded the Distinguished Young Scientist Award of the Czechoslovak Academy of Sciences. During 1984 to 1985 and 1988 he spent a total of 18 months as a research associate at the Oklahoma Medical Research Foundation, Oklahoma City.

Dr. Větvička is the author or co-author of more than 150 original papers and several review articles. He is co-author of the books *Evolution of Immune Reactions, Immunology of Annelids, Aspartic Proteinases, Methods in Cellular Immunology,* and *Evolutionary Mechanisms of Defense Reactions*, and co-editor of the books *Immunological Disorders in Mice* and *Immune System Accessory Cells.* Dr. Větvička's current major research interests include the role of procathepsin D in cancer development, the role of glucan in NK cell activation, and the phylogenic aspects of defense reactions. He holds three patents.

Acknowledgment

We are indebted to Mrs. Jana Větvičková for her valuable comments and indispensable help in the preparation of this book

To Anna, Cristina, Antonio, and Jana

Contents

chapter one

Sources

Contents

I. Preparation of cell suspensions[1]

A. Heat inactivation of fetal calf serum

Materials and reagents

- Fetal calf serum (FCS)
- Thermometer
- Equivalent bottle
- Water bath

Protocol

1. Remove serum from –20°C freezer and allow to thaw overnight in a refrigerator.
2. Fill an equivalent bottle with water and place a thermometer in the water. Fill a water bath to 0.25 in. above the serum line.
3. Preheat the water bath to 56°C. Place the bottles with FCS and a bottle with a thermometer in the water bath.
4. Mix the serum every 10 min to avoid gelling of proteins.
5. When the temperature reaches 56°C, begin timing. Heat inactivate for 30 min with continuing mixing every 10 min.
6. Cool rapidly by placing on ice.
7. Freeze the whole bottles or smaller aliquots.

Comment

The occasional appearance of turbidity or flocculent material does not affect the quality of serum.

B. Bone marrow

Materials and reagents

- RPMI 1640 medium
- Squeeze bottle with 70% ethanol
- Scissors and forceps
- Cutting board or paper towels
- 60 × 15-mm petri dishes
- 3-ml syringes
- 22G1 needles
- 15-ml conical centrifuge tube

Protocol

1. Kill mouse either by cervical dislocation or CO_2 inhalation. Place the mouse on its back on a cutting board (or on paper towel) and soak it with ethanol to reduce the possibility of hair becoming airborne.
2. Make a long transverse cut through the skin in the middle of the abdominal area. Reflect skin from the hindquarters and the hind legs.
3. Separate the legs from the body at the hip joint. Remove the feet. Place the legs in a petri dish containing medium.
4. Remove all muscle tissue from the femur and tibia. Separate femur and tibia and cut off the epiphyses on both ends.
5. Puncture the bone end with a needle and flush out the bone marrow with 3 ml of medium.
6. Remove the large debris and cell clumps by layering the cell suspension over 3 ml of heat-inactivated FCS for 10 min on ice. Wash the suspension by centrifugation at $300 \times g$ for 10 min at 4°C and keep in medium with at least 5% FCS.

Comments

1. For tissue culture studies, perform the entire isolation in a sterile hood and use only sterile instruments, medium, and dishes. Do not use the same scissors you used for cutting the skin for muscle removal.
2. Keep cells on ice.
3. Typical yield of cells is 3 to 7×10^7 per mouse.

C. *Lymph nodes*

Materials and reagents

- RPMI 1640 medium or PBS
- Squeeze bottle with 70% ethanol
- Scissors and forceps
- Cutting board or paper towels
- 60 × 15-mm petri dishes
- Stainless steel screen (Thomas)
- Glass tissue homogenizer (CMS)
- Fetal calf serum (FCS)
- 15-ml conical centrifuge tube

Protocol

1. Kill mouse either by cervical dislocation or CO_2 inhalation. Place the mouse on a cutting board (or on paper towel), and soak it with ethanol to reduce the possibility of hair becoming airborne.

2. Make a cut through the skin in the inguinal region. With fingers on both sides of the cut, pull toward the tail and head until the peritoneal wall is sufficiently exposed.
3. Using the scissors and forceps, remove the draining lymph nodes (inguinal, branchial, and axillary) and separate them from connecting tissues with scissors. Put the lymph nodes into a petri dish containing appropriate medium.
4. Gently tease pieces of tissues over stainless steel screens using the plunger of a 5-ml syringe, or alternatively use a glass homogenizer.
5. Remove large debris and cell clumps by layering the cell suspension over 3 ml of heat-inactivated FCS for 10 min on ice. Wash the suspension by centrifugation at 300 × *g* for 10 min at 4°C and keep in medium with at least 5% FCS.

Comments

1. For tissue culture studies, perform the entire isolation in a sterile hood and use only sterile instruments, medium, and dishes.
2. Keep cells on ice.
3. In addition to these most commonly harvested lymph nodes, cervical nodes (flanking the trachea) and mesenteric nodes (deep in the intestinal mesentery) are also used in immunological experiments.
4. Typical yield from an untreated mouse is 1×10^7 to 5×10^7 cells, depending on the strain used.

D. Spleen

Materials and reagents

- RPMI 1640 medium or PBS
- Squeeze bottle with 70% ethanol
- Scissors and forceps
- Cutting board or paper towels
- 60 × 15-mm petri dishes
- Stainless steel screen (Thomas)
- Glass tissue homogenizer (CMS)
- Fetal calf serum (FCS)
- 15-ml conical centrifuge tube

Protocol

1. Kill mouse either by cervical dislocation or CO_2 inhalation. Place the mouse on a cutting board (or on paper towel) and soak it with ethanol to reduce the possibility of hair becoming airborne.

2. Make a cut through the skin in the inguinal region. With fingers on both sides of the cut, pull toward the tail and head until the peritoneal wall is sufficiently exposed. Soak the peritoneal wall with ethanol.
3. Cut the peritoneal wall, lift the spleen with the forceps, and separate it from connecting tissues with scissors. Put the spleen into a petri dish containing medium.
4. Cut the spleen into several smaller pieces and make a suspension.
5. Gently tease pieces of tissues over stainless steel screens using the plunger of a 5-ml syringe, or alternatively use a glass homogenizer.
6. Remove large debris and cell clumps by layering the cell suspension over 3 ml of heat-inactivated FCS for 10 min on ice. Wash the suspension by centrifugation at 300 × *g* for 10 min at 4°C and keep in medium with at least 5% FCS.

Comments

1. For tissue culture studies, perform the entire isolation in a sterile hood and use only sterile instruments, medium, and dishes. Do not cut the peritoneal wall with the same scissors you used for cutting the skin.
2. Keep cells on ice.
3. The same procedure can be used for lymph nodes or thymus.
4. Typical yield per normal spleen of an untreated mouse is 5×10^7 to 2.5×10^8 cells, depending on the strain used.

E. *Thymus*

Materials and reagents

- RPMI 1640 medium or PBS
- Squeeze bottle with 70% ethanol
- Scissors and forceps
- Cutting board or paper towels
- 60 × 15-mm petri dishes
- Stainless steel screen (Thomas)
- Glass tissue homogenizer (CMS)
- Fetal calf serum (FCS)
- 15-ml conical centrifuge tube

Protocol

1. Kill mouse either by cervical dislocation or CO_2 inhalation. Place the mouse on a cutting board (or on paper towel) and soak it with ethanol to reduce the possibility of hair becoming airborne.
2. Using the scissors, make an incision extending from the xyphoid process to the submandibular region and reflect the skin edges laterally.

3. Make the thoracic incision and separate the edges of the incision.
4. Using the forceps, gently grasp the athymic lobes at the lower pole and lift them.
5. Put the thymus into a petri dish containing medium.
6. Cut the thymus into several smaller pieces and make a suspension.
7. Gently tease pieces of tissues over stainless steel screens using the plunger of a 5-ml syringe, or alternatively use a glass homogenizer.
8. Remove large debris and cell clumps by layering the cell suspension over 3 ml of heat-inactivated FCS for 10 min on ice. Wash the suspension by centrifugation at 300 × *g* for 10 min at 4°C and keep in medium with at least 5% FCS.

F. Peritoneal cells

Materials and reagents

- RPMI 1640 medium or PBS
- Squeeze bottle with 70% ethanol
- Scissors and forceps
- Cutting board or paper towels
- 5-ml syringes
- 22G1 and 23G1 needles
- Thioglycollate medium
- 15-ml conical centrifuge tubes

Protocol

1. Kill mouse either by cervical dislocation or CO_2 inhalation. Place the mouse on its back on a cutting board (or on paper towel) and soak it with ethanol to reduce the possibility of hair becoming airborne.
2. Make a cut through the skin in the inguinal region. With fingers on both sides of the cut, pull toward the tail and head until the peritoneal wall is sufficiently exposed. Soak the peritoneal wall with ethanol.
3. Inject 5 ml of medium into the peritoneal cavity using a 22G1 needle. Gently massage the peritoneum and slowly draw the fluid using a 23G1 needle. You may need to puncture the cavity in several places. Wash the suspension by centrifugation at 300 × *g* for 10 min at 4°C and keep in medium with at least 5% FCS.

Comments

1. The average yield is 2×10^6 cells per mouse. Approximately 30 to 40% of cells are macrophages.
2. You can substantially increase the yield of cells (percentage of macrophages will be more than 90%) by injection of 3 ml of thioglycollate medium i.p. 3 to 6 days prior to harvesting peritoneal cells. The yield

depends on the strain, being between 2×10^7 per mouse (A/J or BALB/c) and 2 to 3×10^7 (C57BL/6, C57BL/10).[2]

3. If you use thioglycollate medium, remember three important points: first, the macrophages are activated, and thus you cannot compare them with macrophages from untreated mice;[2,3] second, always use the same interval between thioglycollate injection and cell harvesting; and third, thioglycollate medium is a bacteriological medium, therefore you must keep it sterile.
4. Several different types of thioglycollate medium exist in the literature. The easiest is to use a prepared medium such as Fluid Thioglycollate Medium from Becton Dickinson. However, for maximal cell yield and up to 95% of macrophages, use the thioglycollate medium, described in Chapter 13.

G. Cell counting

Materials and reagents

- Cell suspension
- Hemocytometer (Fischer)
- Pipettes
- Pasteur pipettes
- Eppendorf tubes
- Türk or trypan blue solution, 0.4% (w/v) in water

Protocol

1. Dilute cells in either Türk or trypan blue solution. Mix thoroughly and add 1 drop into the hemocytometer chamber, using a Pasteur pipette. Do not overfill or underfill the chamber.
2. Count the cells in the 1-mm center square. Count cells on top and left touching middle line of the perimeter of each square.
3. Cell counts/ml = Actual number of cells $\times$ dilution factor $\times 10^4$.

Comments

1. It is possible to increase the accuracy by counting more than one 1-mm square (5 to 10 squares). In that case, substitute *actual number of cells* with *average number of cells per square.*
2. The optimal cell dilution is approximately 20 to 50 cells per square.

1. Cell counting using Türk solution

Materials and reagents

- Cell suspension

- Türk solution
- Hemocytometer
- Pipettes
- Eppendorf tubes

Protocol

Dilute cells in Türk solution. Mix thoroughly and add 1 drop into the hemocytometer chamber using a Pasteur pipette, and then count them.

Comment

Türk solution lyses all erythrocytes, so it is useful for counting cell suspensions prepared from spleen or peripheral blood.

2. Counting of cells in culture plates[4]

Quite often it is necessary to know the exact number of cells in the microtiter plate wells. Cellular ELISA[1,5] requires the accurate quantification of cells involved to obtain reproducible results. The following single-step procedure uses fluorescent dye SYBR green I for direct measurement of cell numbers. SYBR green I binds DNA in preference to RNA and has been shown to bind selectively nuclear DNA in whole cells.[6] The main advantage of this technique is that excitation and emission spectra of this dye are similar to those of fluorescein; therefore, no special equipment is necessary. The dye itself does not fluoresce; thus, there is no need to remove it before measurement.

Materials and reagents

- Cells
- 96-well microtiter plates
- SYBR green I (Molecular Probes, Inc.)
- STORM photoimager (Molecular Dynamics)

Protocol

1. Incubate cells in individual wells of the 96-well microtiter plates, according to the experimental design.
2. Add SYBR green I into each well. Use 5 µm/well of a 1:1000 dilution in water for the live cells, 50 µl/well of a 1:10,000 dilution for formalin-fixed cells.
3. Incubate for 2 h at 37°C.
4. Read the plates using a STORM photoimager.

Comments

1. Cells can be evaluated air dried, formalin fixed, or live.
2. The upper limit of cell numbers is between 2×10^4 to 4×10^4 cells/well, depending on the cell line.

H. Determination of cell viability

1. Trypan blue exclusion

Materials and reagents

- Cell suspension at 2 to 6×10^6 cells/ml
- Trypan blue (Sigma), 0.4% (w/v) in water
- PBS
- 5-ml glass or plastic tube

Protocol

1. Mix 0.25 ml of trypan blue solution, 0.15 ml of PBS, and 0.1 ml of the cell suspension. Allow to stand for 5 to 15 min.
2. Transfer a small amount of the suspension either to the hemocytometer chamber or on cover glass, and count the cells. Nonviable cells will stain blue.

$$\text{Cell viability (\%)} = \frac{\text{number of viable cells}}{\text{number of viable cells} \times \text{number of dead cells}} \times 100$$

Comments

1. Trypan blue has a greater affinity for serum proteins than for cellular proteins. If the background is too dark, resuspend the cells in PBS prior to counting.
2. Do not incubate cells in trypan blue solution longer than 15 min, as viable cells might begin to take up dye.

I. Removal of erythrocytes from cell suspensions

1. Tris-buffered ammonium chloride

Materials and reagents

- Cell suspension
- FCS

- Tris-buffered ammonium chloride
- 15-ml conical centrifuge tube

Protocol

1. Resuspend 1×10^8 cells in 3 ml of Tris-buffered ammonium chloride (working solution) and incubate for 2 min at room temperature.
2. Underlay the cells with 100% FCS and spin at $300 \times g$ for 10 min. Wash two times with washing medium.

Comments

1. Regardless of which method you use, if the cell suspension still looks red, it is necessary to repeat the entire process.
2. The removal of red blood cells is necessary for spleen cell suspension only.
3. Remove red blood cells before counting the cells.

J. Removal of dead cells

1. Centrifugation through FCS

Materials and reagents

- Cell suspension
- FCS
- 15-ml conical centrifuge tube

Protocol

1. Carefully layer 1×10^8 cells in 1 ml of medium over 3 ml FCS in a 15-ml conical centrifuge tube. Centrifuge at $300 \times g$ for 10 min.
2. Discard the supernatant and wash the cells two times in an appropriate medium.

2. Centrifugation through Ficoll-Hypaque[7]

Materials and reagents

- Cell suspension
- Ficoll-Hypaque solution $\varrho = 1.09$ (Sigma; Hypaque-76 from Winthrop Pharmaceuticals)
- RMPI 1640 medium with 5% FCS
- 15-ml conical centrifuge tube

Protocol

1. Add 4 ml of Ficoll-Hypaque solution to the 15-ml conical tube and carefully layer 4 ml of cells (5×10^6 to 1×10^7/ml) over Ficoll-Hypaque (use the wall of the tube).
2. Spin at $2000 \times g$ for 20 min at room temperature with brake off.
3. Collect all of the fluid, add medium with FCS, and centrifuge at $300 \times g$ for 15 min at 4°C. Wash two times in an appropriate medium at $250 \times g$ for 10 min.

Comments

The density of Ficoll-Hypaque mixture is temperature dependent. It is necessary to use the prewarmed Ficoll-Hypaque (20°C) and centrifuge.

II. Isolation of cells

A. Panning[8]

Materials and reagents

- Cell suspension
- Anti-Ig antibody (or anti-CD19, anti-CD20), affinity purified
- PBS
- 15 × 100-mm petri dishes
- RPMI 1640 medium with 5% FCS

Protocol

1. Coat the petri dish with 10 ml of antibody diluted to 50 to 100 µg/ml. Be sure that the entire surface of the dish is submerged. Incubate for 2 h at room temperature or overnight at 4°C.
2. Remove the anti-Ig solution and wash the plate four times with cold PBS.
3. Add cells (up to 2×10^8) in 5 ml of medium with FCS and incubate for 30 min at room temperature.
4. Remove nonadherent cells (T cells).
5. Pour 10 ml of medium with FCS.
6. With a pipette or syringe, draw most of the medium off and strongly pour the medium back into one area of the dish. Repeat several times over new area until the entire dish is covered and most of the cells are removed.

Comments

1. The coated plates can be stored in refrigerator for up to 4 weeks.

2. The anti-Ig solution can be stored and used repeatedly.
3. This method can also be used for isolation of T cells.
4. Check the efficiency of isolation by staining the final suspension with anti-Ig (anti-CD19/20) or anti-CD3 antibodies conjugated to fluorescent marker.

B. *Adherence to Sephadex G-10*[9]

Materials and reagents

- Cell suspension
- RPMI 1640 medium with 5% FCS
- Sephadex G-10 (Pharmacia; Sigma)
- 10- or 50-ml disposable syringe
- 3-way stopcock
- 50-ml conical centrifuge tube
- Nylon wool (Polysciences)
- 1% SDS in PBS
- Pasteur pipette

Protocol

1. Prepare an empty syringe with a closed 3-way stopcock and a needle and prewarm RPMI 1640 medium.
2. Put approximately 10 mg of nylon wool into the bottom of the syringe and press well with the tip of a pipette. Add 5 to 10 ml of medium with FCS and remove all air bubbles. With the tip of a pipette or Pasteur pipette, hold the nylon wool firmly down and open the stopcock to seal the syringe. Close the stopcock.
3. Mix Sephadex G-10 with warm (37°C) medium with FCS.
4. Add 5 to 10 ml of Sephadex G-10 into the syringe and allow to settle down. Open the stopcock, but do not allow all the medium to drain from the Sephadex G-10 mixture. Add the rest of the Sephadex G-10 (leave space for approximately 3 ml of cells). Wash the column with 50 to 150 ml of warm medium with FCS. Close the stopcock.
5. Add the cell suspension (1×10^8) dropwise onto the column, open the stopcock, and allow the cells to penetrate the column. Close the stopcock and incubate for 30 min at 37°C.
6. Open the stopcock and collect the nonadherent cells. Add additional warm medium, but collect only about 10 to 20 ml (depending on the size of the column).
7. Let the possible contaminating Sephadex G-10 settle in a tube (1 to 2 min), decant the medium into a new tube, and wash the cell suspension in medium at $250 \times g$ for 10 min at 4°C.

Comments

1. If you need a sterile cell suspension, perform all steps in a laminar flow hood.
2. In order to deplete the monocytes/macrophages completely, repeat the entire process one more time.
3. Collect the used Sephadex G-10. Wash repeatedly with an excess of water. Pour off the water, add a 1% solution of sodium dodecyl sulfate in PBS, and incubate at room temperature for at least 12 h. Pour off the SDS solution and wash repeatedly with an excess of water (at least 10×). Wash five times with PBS. Resuspend Sephadex G-10 in a small volume of PBS and store at 4°C. Sephadex G-10 can be used up to five times.

C. Purification of cell populations using magnetic microspheres

Materials and reagents

- Magnetic microspheres (Dynal Inc.; Miltenyi Biotec Inc.; Advanced Magnetics)
- Cell suspension
- Appropriate monoclonal antibodies (in case microspheres are not directly coupled to antibodies)
- Centrifuge tubes (size according to the type of magnetic device)
- RPMI 1640 medium
- Plastic Pasteur pipettes
- Rotator
- Magnet

Protocol

1. Incubate cells with monoclonal antibodies in centrifuge tube for 30 min at 4°C on a rotator with end-over-end rotation. The most common concentration is 1 $\mu g/10^6$ cells, but pilot flow cytometry experiment might allow use of even lower concentration.
2. Wash cells two times by centrifugation in RPMI 1640 medium at $150 \times g$ at 4°C.
3. Add anti-Ig-coated magnetic microspheres according to manufacturer's instructions, and incubate for 60 min at 4°C on a rotator with end-over-end rotation.
4. Separate microsphere-coated cells using the magnetic apparatus. After 5 min, carefully transfer the unbound cells to a new tube using a plastic Pasteur pipette, and repeat the magnetic separation.
5. Wash cells two times by centrifugation in RPMI 1640 medium at $150 \times g$ at 4°C and count them.
6. Before the technique becomes routine (and after using a new batch of antibodies), analyze the isolated cell population by flow cytometry.

Comments

1. This technique can be used for depletion (negative selection) or isolation (positive selection) of any cell subset against which the appropriate monoclonal antibodies are available. Thus we will not repeat this method in individual sections of this chapter.
2. You can combine several monoclonal antibodies in one step, e.g., use mouse anti-CD14, mouse anti-CD19, and mouse anti-CD8 and, subsequently, anti-mouse Ig-coated magnetic microspheres for depletion of $CD8^+$ T cells, B cells, and monocytes. The separation will result in pure $CD4^+$ T cells.
3. For best results, repeat isolation steps two times.
4. If you need a sterile cell suspension, perform all steps in a laminar flow hood.
5. Even if optimal results are achieved using magnetic devices available from a company supplying immunobeads, it is not necessary to purchase several rather expensive types of magnets; the use of different magnets is only less convenient.
6. Use of directly labeled magnetic beads allows you to skip Steps 1 and 2 and speeds the entire procedure.
7. Several studies prefer different types of microspheres. However, in our experience, the best results were achieved using the Dynal beads. Microspheres from Miltenyi Biotec Inc. are so small that for most of the experiments (such as flow cytometry) you do not need to remove the beads from the cells, which substantially speeds up the isolation.
8. A modification of this technique is the use of new DETACHaBEAD from Dynal Inc. This product has been developed for rapid detachment of cells bound to Dynabeads. DETACHaBEAD is a special polyclonal antibody preparation that reacts with the Fab fragments of mouse monoclonal antibodies, subsequently disturbing and blocking the interaction between Dynabeads and the cell surface. The whole technique represents a double magnetic isolation and results in a purified population of a selected cell type (i.e., positive separation). The only disadvantage so far is the limited number of Dynabeads available (the new Dynabeads specifically designed for use with DETACHaBEAD are coated with CD4, CD8, and CD19 only), but the manufacturer promises to develop the whole range of Dynabeads.
9. Miltenyi Biotec appears to offer the greatest number of various mixtures of both human and murine antibodies, and different sizes of columns and magnetic devices.

D. *Isolation of monocytes by adhesion and cultivation*

Materials and reagents

- Peripheral blood

- 3.8% sodium citrate
- 24-well tissue culture plates
- 15-ml and 50-ml conical centrifuge tubes
- Ficoll-Histopaque (Sigma)
- 1 *M* $CaCl_2$
- 6% Dextran T-500 (Sigma)
- Saline
- 16 × 125-mm tubes
- Hanks' buffered saline
- RMPI 1640 medium
- Antibiotics
- Trypsin-EDTA solution (BioWhittaker; Intergen)

Protocol

1. Centrifuge freshly drawn peripheral blood with 0.11 ml of 3.8% sodium citrate/ml of blood at 1000 × *g* at room temperature for 20 min.
2. Remove plasma from cells. Place approximately 8 ml of plasma into a new tube and spin for 15 min at 1000 × *g* at room temperature.
3. Add 5 ml of 6% dextran to cells. Bring volume up to 50 ml with saline. Invert several times, loosen cap, and let sit for 30 min at room temperature.
4. Put the remaining plasma into a sterile bottle, and add 20 µl $CaCl_2$/ml of plasma (this is now considered to be an autologous serum). Incubate at 37°C for 1 h.
5. Dilute part of the serum three times with saline.
6. Remove supernatant from the dextran sedimentation tube. Spin at 350 × *g* for 10 min at room temperature. Discard supernatant and save cells.
7. Resuspend these cells in diluted serum (1 to 4 × 10^6/ml). Place 8 ml of cell solution into each 16 × 125-mm tube and underlay with 3 ml of Ficoll-Histopaque. Centrifuge at 1000 × *g* for 25 min at room temperature.
8. Monocytes are in the band at the interface of the gradient. Remove this layer, put the cells into a 50-ml tube, and fill with cold Hanks' buffered saline. Wash two times by centrifugation at 350 × *g* for 10 min at 4°C.
9. Dilute cells to 2 to 3 × 10^6/ml with Hanks' buffered saline. Add 0.1% autologous serum (from Step 4), and place 1 ml of cells/well in a 24-well tissue culture plate. Incubate for 2 h at 37°C.
10. Aspirate nonadherent cells, wash once with RPMI 1640, add 2 ml of RPMI 1640 medium supplemented with 5% autologous serum and antibiotics, and incubate for 3 to 4 days in a humidified 37°C, 5% CO_2 incubator.
11. Resulting monocytes/macrophages can be removed by Trypsin-EDTA solution. (Aspirate medium from wells, add 0.2 ml of Trypsin-EDTA

solution into each well, and incubate for 5 to 10 min at 37°C. Check under the microscope. Pool cells from all wells into a 50-ml tube, add RMPI 1640 medium, and centrifuge at 350 × *g* for 10 min at 4°C).

Comments

1. The resulting population depends on the time of incubation. It is either a mixture of monocytes and macrophages (with a minor contamination of other cell types) or, in the case of 4-day or longer incubations, it consists of macrophages only.
2. If you need a sterile cell suspension, perform all steps in a laminar flow hood.

E. *Isolation of monocytes by adherence*

Materials and reagents

- Peripheral blood mononuclear cells
- RMPI 1640 medium
- 75-cm^2 tissue culture flasks (Corning; Costar)
- 15-ml conical centrifuge tube
- Trypsin-EDTA (BioWhittaker; Intergen) or rubber policeman (Costar)

Protocol

1. Dilute cells in RMPI 1640 medium to a concentration 2 × 10^6/ml.
2. Put 10 ml of diluted cells into a 75-cm^2 tissue culture flask, and incubate 60 min in a humidified 37°C incubator.
3. Aspirate and discard the medium, wash flasks two times with 10 ml of RPMI 1640 medium, and remove adherent cells either by rubber policeman or Trypsin-EDTA treatment.

Comments

1. Some authors recommend using medium supplemented with 10% FCS, human serum, or autologous human serum.[10]
2. If you need a sterile cell suspension, perform all steps in a laminar flow hood.

F. *Removal or purification of monocytes*[11]

Materials and reagents

- Cell suspension
- Baby hamster kidney (BHK) cells (ATCC # 6281)
- RPMI 1640 medium with 10% FCS

- RPMI 1640 medium with 1% FCS
- PBS
- 75-cm^2 tissue culture flasks (Corning; Costar)
- 10 m*M* and 3.3 m*M* EDTA in PBS
- Rubber policeman (Corning; Costar)

Protocol

1. Grow baby hamster kidney cells to confluency in RPMI 1640 medium supplemented with 10% FCS.
2. Remove them from the culture flask with 10 m*M* EDTA-PBS and rinse each flask three times with PBS.
3. Incubate 2×10^7 cells (the cell suspension you want to purify) in 10 ml of RPMI 1640 medium with 1% FCS in the flask where the BHK cells were cultured. Incubate at 37°C for 60 min.
4. Decant the nonadherent cells (i.e., population depleted of monocytes) and rinse the flask three times with PBS. Adherent monocytes are removed after a 10-min incubation with 3.3 m*M* EDTA at 37°C by gentle scraping with a rubber policeman.

Comments

The baby hamster kidney cell-pretreated tissue culture flasks can be stored in a –30°C freezer for up to 6 months.

G. Isolation of fetal liver cells

Materials and reagents

- RPMI 1640 medium
- Squeeze bottle with 70% ethanol
- Scissors and forceps
- Cutting board or paper towels
- 60 × 15-mm petri dishes
- Pasteur pipettes
- 18G1 needles
- 20G1 needles
- 22G1 needles
- Fetal calf serum (FCS)
- 15-ml conical centrifuge tube

Protocol

1. Kill mouse either by cervical dislocation or CO_2 inhalation. Place the mouse on its back on a cutting board (or on paper towel) and soak it with ethanol to reduce the possibility of hair becoming airborne.

2. Make a cut through the skin in the inguinal region. With fingers on both sides of the cut, pull toward the tail and head until the peritoneal wall is sufficiently exposed. Soak the peritoneal wall with ethanol.
3. Cut the peritoneal wall, lift fetuses with the forceps, and separate the wall from connecting tissues with scissors. Put the fetuses into a petri dish containing medium.
4. Remove fetal livers from individual fetuses and immediately dissociate them into single cell suspension by repetitive pipetting with a Pasteur pipette, followed by gentle aspiration through successively smaller hypodermic needles.
5. Remove large debris and cell clumps by layering the cell suspension over 3 ml of heat-inactivated FCS for 10 min on ice. Wash the suspension by centrifugation at $300 \times g$ for 10 min at 4°C and keep in medium with at least 5% FCS.

Comments

1. For tissue culture studies, perform the entire isolation in a sterile hood and use only sterile instruments, medium, and dishes. Do not cut the peritoneal wall with the same scissors you used for cutting the skin.
2. Keep cells on ice.
3. Typical yield per one fetal liver (day 17) is approximately 3 to 4×10^7 cells.
4. Fetuses can be used at 11 to 19 days after mating. The appearance of a vaginal plug is considered day 0.

III. Mitomycin C treatment

Materials and reagents

- Mitomycin C (Sigma), 0.5 mg/ml in PBS
- Cell suspension
- 15-ml conical centrifuge tubes
- Aluminum foil
- PBS
- RPMI 1640 medium supplemented with 10% FCS and antibiotics
- Hemacytometer

Protocol

(*Note:* All materials and reagents must be sterile and proper aseptic technique must be used when handling the cells.)

1. Isolate cells as described elsewhere and prepare cell suspension at concentration of 5×10^7 cells/ml in PBS.
2. Dilute the stock of mitomycin C to 50 μg/ml in PBS.

3. Wrap the tube in aluminum foil and add 100 μl of mitomycin C solution to every 100 μl of cell suspension.
4. Incubate for 20 min at 37°C.
5. Wash the cells six times by centrifugation in RPMI 1640 medium at 150 × *g* at 4°C.
6. Count the cells with a hemacytometer.

Comments

1. Mitomycin C is an antibiotic derived from *Streptomyces caespitosus.* It cross-links DNA strands and thus prevents their separation during replication.
2. Mitomycin C is extremely light sensitive; therefore, it is necessary to prepare a fresh solution each day.
3. Mitomycin C is very toxic.
4. It is crucial to remove all traces of mitomycin C from the cell suspension, as the presence of mitomycin would significantly change the proliferative response of tested population of cells.

IV. Preparation of Mycobacteria *suspension*[12]

Materials and reagents

- 100 mg dried, heat-killed *Mycobacterium tuberculosis* (strain H37Ra)
- 10 ml incomplete Freund's adjuvant (Difco)
- 15-ml plastic tube
- 7-cm-diameter porcelain mortar and pestle
- Gloves

Protocol

1. Put dried *M. tuberculosis* into a mortar.
2. Grind intensively for at least 3 min with the pestle. Color change from gray to white is a good indicator.
3. Add 3 ml of incomplete Freund's adjuvant and continue to grind for an additional 2 min.
4. Transfer the prepared paste into a 15-ml plastic tube.
5. Add 3 ml of incomplete Freund's adjuvant into the mortar, and continue grinding for 30 s. Transfer the paste into the plastic tube.
6. Repeat Step 5.

Comments

1. *M. tuberculosis* may be sensitizing for humans; therefore, always wear protective gloves.

2. Prepared mycobacteria suspension can be stored at 4°C for at least 1 month.

V. Preparation of Listeria *suspension*[13]

Materials and reagents

- *Listeria monocytogenes* (ATCC, strain 15313)
- Tryptose phosphate broth
- 10-ml plastic tubes with caps
- Pasteur pipette
- 37°C incubator
- Water bath
- Refrigerated centrifuge
- PBS

Protocol

1. Grow bacteria to log phase in tryptose phosphate broth in a 37°C incubator.
2. Transfer 1 drop of bacterial suspension into 5 ml tryptose phosphate broth, and incubate overnight in a 37°C incubator.
3. Incubate the tube with bacteria for 60 min in a 70°C waterbath.
4. Wash bacteria by centrifugation 20 min at 900 × *g* at 4°C and discard the supernatant.
5. Resuspend in 10 ml PBS.
6. Repeat Step 4. Resuspend in 1 ml PBS.

VI. ELISA

Enzyme-linked immunosorbent assay (ELISA)[14] is a technique for assaying the presence of antibodies in various fluids. Using any of numerous modifications, this method allows us to qualitatively or quantitatively evaluate a particular antibody activity, to measure an antigen using a defined antibody preparation, or to detect cell-surface antigens. ELISA is one of the most versatile and widely used techniques, and its sensitivity is between 100 pg/ml and 1 ng/ml.

As a general rule, the incubation times used in individual steps might be either overnight at 4°C or 2 h at either room temperature or 37°C. The term *incubate* will be used instead of repeating *overnight at 4°C or 2 h at either room temperature or 37°C* in every step.

It is very difficult to find two laboratories that perform this assay exactly the same way. Readers should try to use the following information as a basis for development of their own modification. One of the more important modifications is the use of 0.25% gelatin or 5% instant milk instead of bovine serum albumin.

Table 1 Various Substrates Used for Visualization of an ELISA

Enzyme	Substrate	OD
Alkaline phosphatase[a]	*p*-Nitrophenyl phosphate	405
	4-Methylumbelliferyl phosphate	365/450
Horseradish peroxidase	*O*-Phenylendiamine	492
	2,2′-Azino-bis(3-ethylbenzthiazoline)	
	Sulfonic acid	414
	O-dianisidine	530
	5-Aminosalicylic acid	474
	3,3′,5,5′-Tetramethylbenzidine	450
β-D-galactosidase	*O*-Nitrophenyl-β-D-galactopyranoside	420
	Chlorophenolic red-β-D-galactopyranoside	574
	Resorufin-β-D-galactopyranoside	570
Urease	Bromcresol purple	588
Acetylcholine esterase	Ellman's reagent	412

Note: Readers seeking more information about various enzyme–substrate combinations should see Reference 15.

[a] The fluorogenic system using 4-methylumbelliferyl phosphate is up to 100 times faster than using *p*-nitrophenyl phosphate.

Various substrates used for visualization of an ELISA assay are listed in Table 1.

A. Indirect ELISA

This technique is used primarily for screening antibodies such as in testing of hybridoma culture supernatants. It does not require the use of preexisting specific antibodies. The wells are coated with the antigen followed by incubation with supernatants. Unbound antibodies are washed out and an Ig-specific antibody conjugated to an enzyme is used for visualization.

Materials and reagents

- ELISA reader
- Antigen
- Test samples
- Enzyme-conjugated antibody
- 96-well U- or flat-bottom microtiter plates (Immulon; Dynatech)
- Substrate
- PBS containing 0.05% Tween 20 (PBS-Tween)
- PBS containing 0.05% Tween 20, 1% BSA, and 0.02% NaN_3 (PBS-BSA-Tween)
- Multichannel pipette
- Carbonate-bicarbonate buffer, pH 9.6
- Plastic squirt bottle
- Paper towels

Protocol

1. Dilute the antigen in carbonate-bicarbonate buffer at a concentration of 5 µg/ml. Using a multichannel pipette, dispense 100 µl of antigen solution into each well on each plate. This coats the plates with antigen.
2. Cover the plates or wrap them in plastic wrap and incubate them.
3. Rinse the plates five times over a sink by filling all wells with PBS-Tween from a plastic squirt bottle. Flick the PBS-Tween into the sink after each rinse.
4. Block residual binding capacity of wells by filling all wells with PBS-BSA-Tween and incubating for 30 min at room temperature.
5. Rinse the plates five times over a sink by filling all wells with PBS-Tween from a plastic squirt bottle. Flick the PBS-Tween into the sink after each rinse. After the last wash, remove residual PBS-Tween by laying the plates face down on paper towels for several seconds.
6. Add 100 µl of test samples diluted in PBS-BSA-Tween to each well. Cover the plates and incubate.
7. Rinse the plates five times over a sink by filling all wells with PBS-Tween from a plastic squirt bottle. Flick the PBS-Tween into the sink after each rinse. After the last wash, remove residual PBS-Tween by laying the plates face down on paper towels for several seconds.
8. Add 100 µl of enzyme-conjugated antibody diluted in PBS-BSA-Tween to each well. Cover the plates and incubate.
9. Rinse the plates three times over a sink by filling all wells with PBS-Tween from a plastic squirt bottle. Flick the PBS-Tween into the sink after each rinse. After the last wash, remove residual PBS-Tween by laying the plates face down on paper towels for several seconds.
10. Add 100 µl of substrate solution to each well, cover the plates, and incubate them for 1 h at room temperature in the dark. Check the color development occasionally. Read the reaction on an ELISA reader using the appropriate filter.

Comments

1. The optimal concentration of enzyme-conjugated antibody should be determined by a pilot experiment utilizing criss-cross serial dilution analysis.
2. Covered antigen-coated plates can be stored at 4°C for several months. Do not let them dry out.
3. The optimal incubation time in the last step depends on the type of enzyme–substrate combination and on the concentration of antibody in tested samples. Stop the reaction (either by reading or by adding a stopping reagent) when positive wells reach the desired intensity.

VII. Incorporation of tritiated thymidine into DNA

As cells enter the "S" phase of the cell cycle, chromosome replication takes place, with the incorporation of soluble nucleotide precursors into newly synthesized DNA. In this assay, dividing cells are pulsed with radioactive [^{3}H] thymidine for several hours, after which the amount of radioactivity incorporated into their DNA is determined by harvesting the cells onto glass fiber filters followed by liquid scintillation counting.

Materials and reagents

- Cells (primary culture, indicator cell lines, etc.)
- Culture medium (depending on the cell culture; for many applications the culture medium of choice is RPMI 1640 medium supplemented with 10% FCS, 100 U/ml penicillin, 100 μg/ml streptomycin, 2 m*M* L-glutamine, and 50 μ*M* 2-mercaptoethanol)
- [^{3}H-methyl] thymidine (approx. 20 to 50 Ci/mmol), sterile (e.g., Amersham; Du Pont NEN)
- 96-well tissue culture plates (flat bottom)
- CO_2 incubator (humidified, set at 37°C and 5% CO_2)
- Multichannel pipette or repeating dispenser
- Cell harvester (Skatron Instruments)
- Glass fiber filter mats (Skatron Instruments)
- Liquid scintillation vials (Research Products International)
- Liquid scintillation cocktail (e.g., Econo-Safe™, Research Products International)
- Liquid scintillation counter

Protocol

1. Plate the indicator cells in 96-well plates in a final volume of 0.1 to 0.2 ml. Normally, each culture condition (e.g., dilution) should be tested in triplicate. Include a background proliferation control in which the cells are cultured in medium alone. Incubate at 37°C in a CO_2 incubator for the appropriate length of time.
2. Dilute the [^{3}H-methyl] thymidine stock solution with culture medium to a concentration of 50 μCi/ml. With the aid of a multichannel pipette, add 20 μl of the diluted solution to each well. This gives a final concentration of 1 μCi per well.
3. Return the plates to the incubator. Incubate for an additional 4 to 18 h. (With rapidly dividing cells, a 4- to 6-h "pulse" gives adequate results.)
4. Harvest the cells onto glass fiber mats using a cell harvester and distilled water.
5. Allow the filters to air dry. Place individual filters into liquid scintillation vials and add 2 to 3 ml of liquid scintillation cocktail.

6. Count in a liquid scintillation counter. Express proliferation data based on cpm or dpm per culture. Alternatively, proliferation data can be expressed as the "Stimulation Index," calculated by dividing the proliferation of the stimulated cells (in cpm) by the background proliferation (in cpm).

Comments

1. The culture medium, the final cell density, and the length of incubation will depend on the nature of the indicator cells and on the nature of the stimulus. Generally, for indicator cell lines (i.e., HT-2, CTLL) which divide rapidly, cell densities of 0.5 to 1×10^5/ml give adequate results. Freshly isolated lymphocytes or mononuclear cells, on the other hand, require higher cell densities (0.5 to 2×10^6/ml). As for incubation times, measurement of proliferation of indicator cell lines to growth-promoting cytokines requires incubations of no more than 48 h, whereas proliferation of lymphocytes to antigens or mitogens may require incubations as long as 72 to 96 h.
2. It is important to use tritiated thymidine of high purity and correct specific activity. It is recommended to use thymidine labeled in the 5-methyl position to avoid the potential risk of labeling RNA due to conversion of thymidine into uridine by demethylation. [^{3}H-methyl] thymidine is commercially available at various specific activities, ranging from 2 to 100 Ci/mmol. The preparations with the lower specific activities are available sterile, in aqueous solution. Ethanol, however, is usually added to the preparations of higher specific activity as a free radical scavenger in order to reduce radiation decomposition. When using these preparations, the investigator should be aware of the final ethanol concentration, as it may be toxic to some cells.
3. Many liquid scintillation formulations are adequate for this purpose; however, because of the relatively large quantities of radioactive waste generated, the use of a biodegradable liquid scintillation cocktail is recommended.
4. Disintegrations per minute (dpm) = counts per minute (cpm) ÷ counting efficiency

VIII. MTT assay

Due to the potential hazards associated with the use of radioactive substances, many investigators prefer the use of nonradioactive detection reagents for the measurement of cellular proliferation. The most common of these reagents is known as the "MTT Assay."[2] Such technique is based on the cleavage of a yellow tetrazolium dye (3-[4,5-dimethylthiazol-2-yl]-2, 5-diphenyl tetrazolium bromide, MTT) into insoluble purple formazan by dehydrogenases in active mitochondria. Dead cells are unable to perform this reaction. In this assay, an

MTT solution is added to the dividing cells and, after a 4-h incubation period, the amount of purple formazan generated is spectrophotometrically determined using a multiwell spectrophotometer or "ELISA reader."

Materials and reagents

- Cells (primary culture, indicator cell lines, etc.)
- Culture medium (depending on the cell culture; for many applications, the culture medium of choice is RPMI 1640 medium supplemented with 10% FCS, 100 U/ml penicillin, 100 μg/ml streptomycin, 2 m*M* L-glutamine, and 50 μ*M* 2-mercaptoethanol)
- 96-well tissue culture plates (flat bottom)
- CO_2 incubator (humidified, set at 37°C and 5% CO_2)
- Multichannel pipette or repeating dispenser
- MTT (Sigma Cat. No. M-5655). Prepare a stock solution of MTT (5 mg/ml) in PBS. Filter.
- Acidic isopropanol (0.04 *N* HCl in isopropanol)
- Multiwell spectrophotometer (ELISA plate reader)

Protocol

1. Plate the cells in 96-well tissue culture plates as described in the previous section, including cells cultured in medium alone as a background proliferation control (triplicates are recommended). Culture in a CO_2 incubator at 37°C for the appropriate length of time.
2. Using a multichannel pipette, add 10 μl of the MTT stock solution per 100 μl of culture medium. (Final MTT concentration is 0.5 mg/ml.)
3. Incubate at 37°C for 4 h.
4. If the cells are nonadherent, add 100 μl/well of the acidic isopropanol solution directly onto the wells. If the cells are adherent, the medium can be discarded first, and then the acidic isopropanol (100 μl per well) can be added directly onto the cells.
5. Mix well and incubate at room temperature until all crystals are dissolved (5 to 10 min).
6. Read on a multiwell spectrophotometer using a test wavelength of 570 nm and a reference wavelength of 630 nm. Read plates within 1 h of addition of the acidic isopropanol.

Comments

1. Red blood cells do not cleave MTT to a significant extent, nor do they interfere with the assay up to concentrations of 2×10^6 cells/ml.[2]
2. Several modifications to the original assay described by Mosmann[2] have been reported. Most of these modifications address the problem of protein precipitation, especially when using a culture medium

containing high serum concentrations, during the formazan-extraction step using acidic isopropanol.[2] Sodium dodecyl sulfate (SDS) appears to minimize protein precipitation but requires longer extraction times, whereas isopropanol accelerates the extraction but leads to increased development of turbidity due to protein precipitation. Extraction reagents in some of these modifications include:[3]

(a) 10% SDS buffered to pH 4.7 with acetate buffer
(b) 10% SDS in 50% isopropanol — 0.01 *N* HCl, pH 5.5
(c) 20% SDS in 50% *N*,N-dimethylformamide buffered to pH 4.7 with acetic acid
(d) 3% SDS in acidified (0.04 *N* HCl) isopropanol

In another modification resulting in improved sensitivity and reproducibility,[7] the normal medium is removed before the addition of MTT. The incubation with MTT is then carried out in serum-free medium devoid of phenol red, thus avoiding potential precipitation of proteins and the need for acidification. The purple formazan product is then dissolved using propanol or ethanol. This assay uses 560 and 690 nm as the test and reference wavelengths, respectively.

3. Another nonradioactive detection assay for cell proliferation is available from Amersham. This assay is based on the incorporation of 5-bromo-2′ deoxyuridine (BrdU), a thymidine analog, into DNA, followed by detection of incorporated BrdU using a specific peroxidase-labeled monoclonal antibody against BrdU.

IX. ^{51}Cr labeling

Materials and reagents

- Target cells
- $Na_2{}^{51}CrO_4$, 5 mCi/ml, specific activity 250 to 500 mCi/mg (Amersham)
- RPMI 1640 medium with 25 m*M* HEPES and 10% FCS
- 15-ml conical centrifuge tubes
- Tubes suitable for your γ-counter
- γ-Counter

Protocol

1. Resuspend cells in medium at a concentration 1×10^7/ml. Check the viability; use only cells with viability higher than 95%.
2. Add $Na_2{}^{51}CrO_4$ to the cells (500 µCi per 1×10^7 cells) and incubate for 60 min at 37°C with occasional shaking.
3. Wash the cells at least five times by centrifugation at $300 \times g$ for 10 min at room temperature. Check the radioactivity in the supernatant from

last centrifugation. If still significantly higher than a background, repeat the washing procedure.

Comment

A major disadvantage of this very common technique is the rather high spontaneous ^{51}Cr release from target cells, especially when the cells are grown in monolayers and need to be trypsinized in order to obtain a single cell suspension prior to labeling.

References

1. Fernandez-Botran, R. and Větvička, V., *Reagents in Cellular Immunology*, CRC Press, Boca Raton, 1995.
2. Mosmann, T., Rapid colorimetric assay for cellular growth and survival: application to proliferation and cytotoxicity assays, *J. Immunol. Reagents*, 65, 55, 1983.
3. Niks, M. and Otto, M., Towards an optimized MTT assay, *J. Immunol. Reagents*, 130, 149, 1990.
4. Myers, M. A., Direct measurement of cell numbers in microtiter plate cultures using the fluorescent dye SYBR green I, *J. Immunol. Meth.*, 212, 99, 1998.
5. Feit, C., Bartal, A. H., Tayber, G., Dymbort, G., and Hirshaut, Y., An enzyme-linked immunosorbent assay (ELISA) for the detection of monoclonal antibodies recognizing antigens expressed on viable cells, *J. Immunol. Reagents*, 58, 301, 1983.
6. Suzuki, T., Fujikura, K., Higashiyama, T., and Takata, K., DNA staining for fluorescence and laser confocal microscopy, *J. Histochem. Cytochem.*, 45, 49, 1997.
7. Denizot, F. and Lang, R., Rapid colorimetric assay for cell growth and survival. Modifications to the tetrazolium dye procedure giving improved sensitivity and reliability, *J. Immunol. Reagents*, 89, 271, 1986.
8. Wysocki, W. L. and Sato, V. L., "Panning" for lymphocytes: a method for cell selection, *Proc. Natl. Acad. Sci. U.S.A.*, 75, 2844, 1978.
9. Ly, I. A. and Mishell, R. I., Separation of mouse spleen cells by passage through columns of Sephadex G-10, *J. Immunol. Reagents*, 5, 239, 1974.
10. Edelson, P. J. and Cohn, Z. A., Purification and cultivation of monocytes and macrophages, in *In vitro Reagents in Cell-Mediated and Tumor Immunity*, Bloom, B. A. and Davis, J. R., Eds., Academic Press, San Diego, 1976, 333.
11. Ackerman, S. K. and Douglas, S. D., Purification of human monocytes on microexudate-coated surfaces, *J. Immunol.*, 120, 1372, 1978.
12. van Eden, W., Wagenaar-Hilbers, J. P. A., and Wauben, M. H. M., Adjuvant arthritis in the rat, in *Current Protocols in Immunology*, Coligan, J. E., Kruisbeek, A. M., Margulies, D. H., Shevach, E. M., and Strober, W., Eds., John Wiley & Sons, New York, 1998, 15.4.1.
13. Campbell, P. A., Canono, B. P., and Drevets, D. A., Measurement of bacterial ingestion and killing by macrophages, in *Current Protocols in Immunology*, Coligan, J. E., Kruisbeek, A. M., Margulies, D. H., Shevach, E. M., and Strober, W., Eds., John Wiley & Sons, New York, 1998, 14.6.1.

14. Engvall, E. and Perlman, P., Enzyme-linked immunosorbent assay (ELISA): quantitative assay of immunoglobulin G, *Immunochemistry,* 8, 871, 1971.
15. Porstman, B. and Porstman, T., Chromogenic substrates for enzyme immunoassays, in *Nonisotopic Immunoassays,* Ngo, T. T., Ed., Plenum Press, New York, 1988, 57.

chapter two

Isolation and characterization of neutrophils

Contents

Neutrophils or polymorphonuclear leukocytes (PMNs) are the most abundant type of leukocyte in normal peripheral blood. They participate in the effector phase of immune responses, playing important roles in inflammation and in the pathogenesis of a number of diseases.[1,2] Neutrophils are phagocytic cells and are the major cell population involved in acute inflammatory responses. Furthermore, they migrate to, and accumulate at, sites of complement activation. Neutrophils can be isolated in high numbers and with relatively high purity from the blood of humans and other animals.[3,4] Most conventional isolation procedures have been based on the separation of neutrophils from mononuclear cells based on differences in cell density, using centrifugation through Percoll and/or Ficoll-Paque gradients.[3,4] More recently, protocols based on immunomagnetic separation using anti-CD15 beads have been reported to yield preparations of improved purity, compared to the density techniques.[5,6] It should be pointed out that neutrophils are generally considered fragile cells, and are easily damaged if handled improperly. In addition, neutrophils can be potentially activated as a consequence of the isolation techniques or the presence of bacterial contaminants in the isolation reagents or equipment. Therefore, great care should be exercised throughout. In this chapter we describe both density- and immunomagnetic-based isolation procedures. These techniques have been reported to produce neutrophils of comparable functional attributes.[5,6]

I. Percoll sedimentation and Ficoll-Paque centrifugation

Materials and reagents

- Anticoagulated blood (heparin, EDTA, citrate)
- Dextran T-500 (Amersham Pharmacia Biotech), 3% solution in 0.85% NaCl
- Ficoll-Paque or Histopaque, density 1.077 g/ml. (Amersham Pharmacia Biotech or Sigma)
- Saline solution (0.85% NaCl)
- 0.2% and 1.6% NaCl solutions
- Phosphate-buffered saline containing 10 m*M* D-glucose (PBS-glucose)
- 50-ml conical centrifuge tubes
- Graduated plastic cylinder
- 10-ml syringe and tubing
- Refrigerated centrifuge

Protocol

1. Obtain blood by venipuncture (100 to 200 ml).
2. Mix blood with an equal volume of 3% dextran T-500 solution (room temperature). Transfer to a graduated plastic cylinder or 50-ml centrifuge tubes. Place in upright position and allow erythrocytes to sediment for approximately 20 min at room temperature.
3. After a clean interface between the leukocyte-rich plasma and erythrocytes has formed, carefully aspirate the upper layer (plasma) to a centrifuge tube and pellet cells by centrifuging at 250 × *g* for 10 min at 4°C. Discard supernatant. This layer will still contain contaminating erythrocytes, but these will be removed later.
4. Resuspend the cell pellet in a volume of saline solution equal to the starting volume of blood.
5. Transfer approximately 40 ml of the cell suspension into a 50-ml conical centrifuge tube. Carefully underlayer with 10 ml of Ficoll-Paque solution, using a pipette or syringe with attached tubing. Centrifuge at 400 × *g* for 40 min at 20 °C.
6. Aspirate both the top saline (containing the mononuclear cells) and Ficoll-Paque layers, leaving the neutrophil pellet.
7. To remove contaminating erythrocytes, resuspend the neutrophil pellet in 20 ml of ice-cold 0.2% NaCl solution for 30 s. Immediately restore isotonicity by adding 20 ml of ice-cold 1.6% NaCl.

8. Centrifuge at 250 × *g* for 10 min at 4°C and discard supernatant. The isotonic lysis procedure may be repeated until most erythrocytes are removed.
9. Resuspend cell pellets in ice-cold PBS-glucose. Combine tubes and count. Adjust to desired cell density with additional PBS-glucose solution.

Comments

1. Viability should be > 95%.
2. When counting cells, the use of 3% acetic acid diluent with or without crystal violet allows visualization of cell morphology and differentiation of neutrophils from contaminating leukocytes.
3. If mononuclear cells are to be isolated, it is recommended to perform first the Ficoll-Paque centrifugation, followed by the dextran sedimentation and RBC lysis.
4. Even though sterility is not essential due to the short incubation periods, the use of sterile equipment and solutions is recommended, as neutrophils may be activated by bacterial contaminants.

II. Magnetic cell sorting with anti-CD15 beads

Although CD15 is not exclusively expressed by neutrophils (it is also expressed in some eosinophils and some myeloid precursors cells), immunomagnetic isolation of human neutrophils from peripheral blood with anti-CD15 beads has been reported to lead to highly enriched (>99%) neutrophil preparations. The procedure described by Zahler et al.[5] is given below.

Materials and reagents

- Anticoagulated blood (EDTA)
- HBSS (Ca^{2+} and Mg^{2+} free)
- HBSS (Ca^{2+} and Mg^{2+} free) containing 2% FBS (HBSS-2% FBS)
- Gelatin, 0.1% solution in HBSS (Ca^{2+} and Mg^{2+} free)
- 10X HBSS (Ca^{2+} and Mg^{2+} free)
- Percoll, diluted in HBSS to a density of 1.090 g/ml (Amersham Pharmacia Biotech)
- Anti-CD15 immunomagnetic beads (MACS CD15 Microbeads, Miltenyi Biotec)
- Diff-Quik or Wright-Giemsa stain (Fisher)
- Trypan blue stain
- Refrigerated centrifuge
- Magnetic separation apparatus (MACS™, Miltenyi Biotec)
- Positive selection MACS column (type MS^+/RS^+ for up to 10^7 positive cells or LS^+/VS^+ for up to 10^8 positive cells (Miltenyi Biotec)
- Platform rocker

- 50-ml conical centrifuge tubes
- Pipettes

Protocol

1. Perform venipuncture and obtain blood.
2. Centrifuge at 400 × *g* for 15 min. Discard the platelet-rich plasma and collect the "buffy coat."
3. Centrifuge and resuspend cell pellet in 50 μl HBSS-2% FBS buffer. Add 50 μl of anti-CD15 immunomagnetic beads (manufacturer recommends 20 μl of beads per 10^7 total cells). Incubate for 15 min at 6 to 12°C on a platform rocker.
4. Prepare the magnetic column, load cells, and perform isolation according to the manufacturer's recommendations. Collect bound cells (neutrophils).
5. Centrifuge at 250 × *g* for 10 min at 4°C and discard supernatant.
6. Resuspend cell pellets in ice-cold PBS-glucose. Count and assess viability.

Comments

1. Expect approximately a purity of >99% neutrophils and a yield of 0.5×10^6 cells/ml of blood.
2. Functionally, neutrophils isolated by the immunomagnetic procedure have been found to be comparable to neutrophils isolated by dextran sedimentation and Ficoll-Paque centrifugation method.[5]

III. Functional assessment

A. Measurement of phagocytic activity

The phagocytic activity of neutrophils can be measured using a variety of microorganisms, such as bacteria (e.g., *Staphylococcus aureus, Escherichia coli*) or yeasts (*Candida albicans*).[3–5] Detection of phagocytized organisms can be accomplished either through staining and microscopical analysis or by labeling of the microbial cell suspension with either fluorochromes or radioactive precursors (e.g., amino acids, uracil) and then measuring the cell-associated fluorescence or radioactivity, respectively. The procedure described here utilizes ^{14}C heat-killed *Staphylococcus aureus*.[4]

Materials and reagents

- Neutrophil suspension (5×10^6 per ml) in HBSS
- Fresh autologous or AB human serum
- HBSS (with Ca^{2+} and Mg^{2+})
- ^{14}C heat-killed *S. aureus* (strain 502A, ATCC 27217)

- Stop solution (HBSS containing 10% FCS and 2 m*M* NaF)
- Saline (0.85% NaCl)
- 75 × 10-mm capped tubes
- Refrigerated centrifuge
- 56°C water bath
- Scintillation vials and fluid
- Scintillation counter

Protocol

1. Prepare enough 75 × 10-mm tubes to measure phagocytosis at different time points (e.g., 0, 5, 10, 20, and 30 min). Run each sample in duplicate.
2. Set tubes on ice. Add to each tube 0.5 ml neutrophil suspension, 0.1 ml serum, and 0.3 ml HBSS. Keep the "0 time" control tubes on ice. Incubate the remaining tubes at 37°C for 5 min.
3. Add 0.1 ml of the labeled, heat-killed *S. aureus* to each tube. Cap and place on a rotator.
4. At the end of each time point, remove tubes from the rotator and stop phagocytosis by adding 2 ml of ice-cold stop solution. Mix and place on ice.
5. Add 2 ml of stop solution and 0.1 ml of bacterial suspension to the "0 time" controls.
6. Centrifuge tubes at 150 × *g* for 5 min at 4°C. Discard the supernatant.
7. Wash pellets two times by resuspending in 3 ml ice-cold stop solution, centrifuging at 150 × *g* for 5 min at 4°C and discarding the supernatant. Avoid vortexing the cells, as this may break them and release the bacteria.
8. Add 0.5 ml of 0.5 *M* NaOH solution to each tube. Mix by vortexing and incubate 3 h at 56°C.
9. Control tubes should be set at this point by mixing 0.1 ml of labeled bacteria and 0.5 ml of 0.5 *M* NaOH solution. Incubate at least 15 min at 37°C.
10. Add 0.2 ml of 3% acetic acid solution and 0.5 ml of saline to each tube. Control tubes get 0.2 ml of 3% acetic acid and 0.4 ml saline.
11. Remove 1 ml aliquots from each tube and transfer to scintillation vials. Add 10 ml of scintillation fluid to each vial. Count in scintillation counter.
12. Calculate the percent phagocytosis according to the formula:

$$\%\ \text{phagocytosis} = (\text{sample cpm}) \times 100 \div (\text{control cpm})$$

Comments

The labeled bacteria suspension is prepared by growing *S. aureus* 502A (ATCC) in Trypticase soy broth (TSB) containing 10 μCi/ml of a ^{14}C-labeled amino acid or uracil for approximately 4 h at 37 °C in a shaking incubator. The bacteria are then centrifuged at $1000 \times g$ for 10 min at 4°C and washed two times in HBSS. Count and adjust bacteria to 2.5×10^8 per ml in HBSS. Heat kill bacteria by inoculating 20 ml of TSB and incubating overnight at 37°C and then boiling the suspension for 45 min. Centrifuge and wash as above. Count and adjust bacteria to 2.5×10^8 per ml in HBSS. Store at –20°C.

B. *Microbicidal assay*

Materials and reagents

- Neutrophil suspension (5×10^6 per ml) in HBSS
- Fresh autologous or AB human serum
- HBSS (with Ca^{2+} and Mg^{2+})
- Live *Staphylococcus aureus* (2.5×10^8/ml in HBSS, prepared as in the previous protocol)
- Trypticase soy agar (TSA)
- Sterile water
- 15-ml glass tubes (sterile)
- 100 × 15-mm petri dishes
- Centrifuge

Protocol

1. Prepare enough 15-ml glass tubes to examine bacterial killing at various time points (0 to 90 min). Run each sample in duplicate.
2. Into each tube add 0.5 ml neutrophil suspension, 0.1 ml serum, and 0.3 ml HBSS. Mix and incubate 5 min at 37°C.
3. Prepare neutrophil-free controls by substituting 0.1 ml HBSS for the 0.5 ml of the neutrophil suspension.
4. Add 0.1 ml of live *S. aureus* suspension to each tube, including controls. Place on a rotator and incubate for 20 min at 37°C.
5. At different time points (20 to 90 min), centrifuge the corresponding tubes at $150 \times g$ for 5 min at 4°C in order to pellet the cells and phagocytosed bacteria. Remove the supernatant by decantation and set it aside. Save the pellet as well.
6. Transfer 0.1 ml of each supernatant to tubes containing 10 ml of sterile water. Mix by vortexing. Seed TSA plates with 1.0 and 0.1 ml from each of the tubes. These samples measure free bacteria.
7. In the meantime, resuspend pellets in residual supernatant and mix by vortexing. Take a 10-µl aliquot and dilute in 10 ml of sterile water. Vortex vigorously and incubate for 10 min in order to lyse the cells.
8. Plate 1.0, 0.1, and 0.01 ml aliquots (final 10^{-4}, 10^{-5}, and 10^{-6} dilutions). These cultures measure cell-associated bacteria.
9. Incubate agar plates overnight at 37°C.
10. Count colonies in plates having less than 500 colonies.

11. Calculate the following parameters:
    (a) Number of extracellular bacteria (Step 6)
    (b) Number of bacteria surviving: Number of extracellular bacteria (a) + number of bacteria in cell pellet (Step 8)
    (c) Total number of cell-associated bacteria: Number of bacteria in initial inoculum – extracellular bacteria (a)
    (d) Number of cell-associated bacteria killed: Total cell-associated bacteria (c) – number of surviving bacteria associated with cells (Step 8)
    (e) Total number of bacteria killed: Number of bacteria in initial inoculum – total number that survived (b)

C. *Nitroblue tetrazolium dye reduction test*

The ability of neutrophils to reduce nitroblue tetrazolium (NBT) is a measure of their NADPH oxidase activity and their ability to generate reactive oxygen intermediates.[4,7] Superoxide (O_2^-) and other products of the respiratory burst reduce NBT to formazan, a blue-black precipitate that can be determined spectrophotometrically. Alternatively, the neutrophils can be fixed and examined microscopically after incubation with NBT, allowing determination of the percentage of cells containing reduced dye (formazan). The spectrophotometric procedure described here is a modification of the technique reported by Jarstrand et al.[7] The slide technique has been described by Absolom.[8]

Materials and reagents

- Neutrophil suspension (1×10^7/ml in HBSS)
- HBSS
- Fresh autologous or AB human serum
- Nitroblue tetrazolium (Sigma, N-6876). Prepare a 0.1% w/v in HBSS the same day of the assay.
- 0.14 *M* HCl
- Dimethylsulfoxide (DMSO)
- Stimuli (stock solutions):
    (a) Lipopolysaccharide (LPS, 100 μg/ml) (e.g., Sigma L-3012)
    (b) Formyl-methionyl-leucyl-phenylalanine (FMLP, Sigma F-3506). Prepare a 10-m*M* stock solution in DMSO and freeze at –80°C.
    (c) Phorbol myristate acetate (PMA, Sigma P-8139). Prepare a 5-mg/ml stock solution in DMSO and store at –80°C.
    (d) A23187 Calcium Ionophore (Sigma C-7522). Prepare a 1-mg/ml stock solution in DMSO and freeze at –80°C.
    (e) Serum-treated Zymosan (STZ).[9] Resuspend Zymosan (Sigma Z-4250) to a concentration of 5 mg/ml in HBSS, boil for 10 min, and wash two times with HBSS. Opsonize the washed Zymosan particles by incubating in the presence of fresh human AB serum (0.5 ml serum per 2.5 g of Zymosan) for 30 min at 37°C and wash

twice in cold HBSS. Resuspend in HBSS to 5 mg/ml and use immediately.

- 10 × 75-mm tubes
- Water bath (70°C)
- Spectrophotometer (580 nm)

Protocol (spectrophotometric method)

1. In test tubes mix 0.2 ml of the neutrophil suspension and 0.15 ml of serum. Pre-incubate at 37°C for 5 min.
2. Add 0.25 ml of NBT solution to each tube. Immediately add 6 μl of HBSS (unstimulated controls) or of the desired 100× stimuli solution in HBSS (e.g., PMA). Incubate for 30 min at 37°C with constant agitation.
2. Stop the reaction by adding 1 ml of 0.14 *M* HCl solution.
3. Centrifuge at 250 × *g* for 10 min at room temperature. Decant supernatant.
4. Add 1.5 ml of DMSO to the pellet and place in a 70°C water bath for 60 min with intermittent agitation.
5. Read OD at 580 nm.

Additional materials and reagents (slide method)

- HBSS containing 5% bovine serum albumin (HBSS-BSA)
- Methanol
- Safranin or Wright-Giemsa stain
- Microscope slides
- Cytocentrifuge

Protocol (slide method)

1. In test tubes, mix 10 μl of the neutrophil suspension (1×10^5 cells), 0.25 ml of NBT solution, 5 μl of HBSS, or 100× stimuli solution and enough HBSS-BSA to bring the volume to 0.5 ml.
2. Incubate at 37°C for 30 min.
3. Transfer contents of the tubes to a slide using a cytocentrifuge.
4. Allow slides to air dry and fix in absolute methanol for 1 min.
5. Counterstain with safranin (1 min) or freshly filtered Wright-Giemsa stain.
6. Examine microscopically under oil immersion and record the fraction of cells containing formazan (purple granules). Examine at least 200 cells.

D. Superoxide anion (O_2^-) generation

Generation of O_2^- under unstimulated and stimulated conditions can be assayed by measuring the superoxide dismutase (SOD)-inhibitable reduction of ferricytochrome *c*, as described by Pick and Mizel.[10]

Materials and reagents

- Neutrophil suspension (1×10^6/ml in HBSS)
- HBSS containing 0.1% gelatin
- Fe^{3+} Cytochrome *c*, 10 m*M* solution (e.g., horse heart, Type VI, Sigma C-7752)
- Superoxide dismutase, 5 mg/ml (SOD; bovine erythrocytes, Sigma S-2515)
- Stimuli (stock solutions) (see above protocol)
- 96-well plates
- Multi-well spectrophotometer (550 nm)
- Incubator (37°C)
- Pipettes

Protocol

1. Plate neutrophils (1×10^5/well) in a final volume of 200 µl of HBSS containing 0.1% gelatin, and 100 µ*M* cytochrome *c*.
2. Initiate the reaction by adding the desired stimulus. For example, PMA (0 to 10 ng/ml), LPS (0.1 to 1 µg/ml), FMLP (0 to 10 µ*M*), or STZ (0 to 0.5 mg/ml). Reactions should be performed in duplicates and against identical control wells containing 20 µg/ml SOD.
3. Reactions should also include controls for spontaneous release (unstimulated cells).
4. Mix and immediately measure absorbance in a multi-well spectrophotometer at 550 nm.
5. Read absorbances every 10 min for 1 h, incubating the plate at 37°C between readings.
6. Calculate O_2^- generation using an extinction coefficient of 21.1×10^{-3} *M*/cm for reduced cytochrome *c*. Results are expressed as nanomoles of cytochrome *c* reduced/5×10^5 cells/30 min, after subtraction of the SOD and spontaneous release controls.

E. Hydrogen peroxide formation

The generation of hydrogen peroxide (H_2O_2) by neutrophils can be measured based on the oxidation of homovanillic acid (HVA) into a fluorescent dimer, a reaction that is catalyzed by horseradish peroxidase (HRPO) and depends on the H_2O_2 generated by the cells. The procedure described here is adapted from that of Ruch et al.[11] Both end-point and kinetic[4] techniques have been applied to this method.

Materials and reagents

- Neutrophil suspension (1×10^7/ml in PBS-glucose)

- Phosphate-buffered saline-glucose (PBS-glucose), containing 138 m*M* NaCl, 2.7 m*M* KCl, 0.6 m*M* $CaCl_2$, 1.0 m*M* $MgCl_2$, 5 m*M* D-glucose, 10 m*M* NaH_2PO_4/Na_2HPO_4, pH 7.4.
- HBSS
- H_2O_2 solution (1 m*M*, freshly diluted in water from 30% stock)
- Homovanillic acid, 10 m*M* in HBSS (Sigma H-1252)
- Horseradish peroxidase, 100 U/ml in HBSS (Type II, Sigma P-8250)
- Superoxide dismutase, 5 mg/ml in HBSS (SOD; bovine erythrocytes, Sigma S-2515)
- Sodium azide, 10 m*M* in HBSS
- Stop solution: 100 m*M* glycine-NaOH buffer containing 25 m*M* EDTA, pH 12.0
- Stimuli (100×): PMA (10 μg/ml), FMLP (10 μ*M*), STZ (50 mg/ml) (see above)
- Centrifuge
- Cuvettes
- Spectrofluorometer, set at 312 nm excitation and 420 nm emission (preferably with temperature control and sample stirring)
- Centrifuge
- 12 × 75-mm tubes
- Pipettes

Protocol (end-point method)

1. In 12 × 75-mm tubes, mix 1.7 ml of prewarmed (37°C) PBS-glucose, 20 μl each of HVA, HRPO, SOD, and azide solutions, and 0.2 ml of the neutrophil suspension. Set blank tubes in which the neutrophil suspension is substituted by 0.2 ml PBS-glucose.
2. Place tubes in a water bath at 37°C; allow 2 min for the temperature to equilibrate.
3. Start reaction by addition of 20 μl of (100×) stimuli solutions. Nonstimulated controls receive 20 μl PBS-glucose.
4. Incubate for 30 min at 37°C with occasional mixing. Stop the reaction by adding 0.25 ml of the stopping solution.
5. Centrifuge tubes at 1200 × *g* for 10 min at room temperature.
6. Measure fluorescence in the clear supernatant. The fluorophore is stable for at least 1 to 2 h at room temperature.
7. To construct a standard curve, substitute 0.2 ml of different H_2O_2 dilutions (0.5 to 10 μ*M* H_2O_2, final) instead of the neutrophil suspension.

Protocol (kinetic method)

1. First, create standard H_2O_2 curve. Mix 1.94 ml of PBS-glucose, 20 μl each of HVA, HRPO, and H_2O_2 solutions in a sample cuvette (H_2O_2 concentrations range from 0.5 to 10 μ*M*).

2. Incubate for 5 min at 37°C. Determine the fluorescence of each standard and construct standard curve.
3. For the samples, mix 1.7 ml of assay buffer, 20 μl each of HVA, HRPO, SOD, and azide solutions, and 0.2 ml of the neutrophil suspension in a clean cuvette.
4. Place cuvette in spectrofluorometer and begin stirring. After allowing approximately 2 min for the temperature to equilibrate (37°C), record baseline fluorescence.
5. Add 20 μl of the stimulus solution and begin recording fluorescence.
6. Calculate lag time (the time from stimulus addition to intercept of the baseline with the extrapolated line of maximum slope).
7. Calculate maximum rate of H_2O_2 formation (rate is calculated by relating the fluorescence change to standard curve obtained in Step 2). For example, if the maximum rate was obtained between times of t_1 and t_2 min, and these correspond in the standard curve to readings of x and y μmol H_2O_2/l, respectively:

$$\left[t_2(\mu\text{mol } H_2O_2/l) - t_1(\mu\text{mol } H_2O_2/l) \times 2 \text{ ml} \times 1000 \text{ nmol}/\mu\text{mol}\right]$$

$$\div 1000 \text{ ml/l} = \text{nmol } H_2O_2/\text{min}/2 \times 10^6 \text{ cells}$$

References

1. Fantone, J. C. and Ward, P. A., Role of oxygen-derived free radicals and metabolites in leukocyte dependent inflammatory reactions, *Am. J. Pathol.*, 107, 397, 1982.
2. Weiss, S. J., Tissue destruction by neutrophils, *New Eng. J. Med.*, 320, 365, 1989.
3. Glasser, L. and Fiederlein, R. L., The effects of various cell separation procedures on assays of neutrophil function, *Am. J. Clin. Pathol.*, 93, 662, 1990.
4. Clark, R. A. and Nauseef, W. M., Preparation and functional analysis of other human lymphoid and non lymphoid cells, in *Current Protocols in Immunology*, Coligan, J.E., Kruisbeek, A. M., Margulies, D. H., Shevach, E. M., and Strober, W., Eds., Greene Publishing and Wiley Interscience, New York, 1991, 7.23.1.
5. Zahler, S., Kowalski, C., Brosig, A., Kupatt, C., Becker, B. F., and Gerlach, E., The function of neutrophils isolated by a magnetic cell separation technique is not altered in comparison to a density gradient centrifugation method, *J. Immunol. Reagents*, 200, 173, 1997.
6. Soltys, J., Swain, S. D., Sipes, K. M., Nelson, L. K., Hanson, A. J., Kantele, J. M., Jutila, M. A., and Quinn, M. T., Isolation of bovine neutrophils with biomagnetic beads: comparison with standard Percoll density gradient isolation reagents, *J. Immunol. Reagents*, 226, 71, 1999.
7. Jarstrand, C., Lahnborg, G., and Berghem, L., Human granulocyte function during dextran infusion, *Acta Chir. Scand.*, 489 (Suppl.), 279, 1979.
8. Absolom, D. R., Basic reagents for the study of phagocytosis, in *Reagents in Enzymology*, DiSabato, G. and Everse, J., Eds., Academic Press, Orlando, 132, 95, 1986.

9. Sedgwick, J. B., Shikama, Y., Nagata, M., Brener, K., and Busse, W. W., Effect of isolation protocol on eosinophil function: Percoll gradients versus immunomagnetic beads, *J. Immunol. Reagents,* 198, 15, 1996.
10. Pick, E. and Mizel, D., Rapid microassays for the measurement of superoxide and hydrogen peroxide production by macrophages in culture using an automatic enzyme immunoassay reader, *J. Immunol. Reagents,* 46, 211, 1981.
11. Ruch, W., Cooper, P. H., and Baggiolini, M., Assay of H_2O_2 production by macrophages and neutrophils with homovanillic acid and horseradish peroxidase, *J. Immunol. Reagents,* 63, 347, 1983.

chapter three

Isolation and characterization of eosinophils

Contents

Eosinophils are bone marrow–derived granulocytes whose granules contain basic proteins that are stained with acidic dyes, such as eosin. Eosinophils comprise approximately 2 to 5% of blood leukocytes in healthy, nonallergic individuals. Although these cells are capable of phagocytosis, they are thought to function mainly in the defense against certain types of infectious agents, including helminths. In addition, eosinophils also appear to play a significant role in the inflammation and tissue injury following immediate hypersensitivity reactions.[1,2] The production and activation of eosinophils are under the control of T cells, particularly the $CD4^+$ Th2 subset, responsible for the production of IL-5, the major eosinophil activating factor.[1,2] After activation, resting eosinophils increase their size and decrease their density.[3,4] Activated eosinophils are more potent at mediating antibody-dependent cell cytotoxicity (ADCC) and producing a variety of inflammatory mediators, including leukotriene C_4 (LTC_4) and O_2^-.[4,5]

The low number of eosinophils in the peripheral blood of normal individuals has made it relatively difficult to obtain eosinophil preparations of adequate numbers and high purity for functional studies. Protocols based on the density separation of eosinophils from neutrophils, the main contaminant cell type, have been used for years. Although these techniques result in eosinophil preparations of high viability, they have the disadvantage of relatively low yield and sometimes variable purity. Moreover, most of the isolated eosinophils are of "high-density" (1.095 to 1.100 g/ml) and presumably represent nonactivated cells.[6,7] In recent years, a negative selection protocol based on the removal of neutrophils through magnetic cell sorting with anti-CD16-coated beads has been shown to significantly improve yield and purity over density-gradient techniques.[8] Preparations obtained by magnetic cell sorting differ from density-based procedures in that "hypodense eosinophils" are also present.[6,7] Furthermore, functional differences between density or magnetically sorted eosinophils have been observed. For example, Casale et al.[7] reported that eosinophils prepared by magnetic-sorting protocols are less responsive to lipid chemoattractants (LTC_4, platelet-activating factor [PAF]) than preparations obtained by Percoll-gradient centrifugation. Sedgwick et al.[6] demonstrated that although magnetic- and density-sorted eosinophils were similar in their *in vitro* survival and adhesion, the former preparations displayed increased production of LTC_4, expontaneous O_2^- generation, and expression of activation markers, such as CD18 and CD54 (ICAM-1). It is not yet clear whether these functional differences are reflections of the presence of a population of less dense eosinophils in the magnetically sorted preparations or are potential consequences of partial eosinophil activation by-products released by neutrophils upon binding to the anti-CD16 beads. These differences should be kept in mind when considering the isolation procedure to be used for a particular experiment. Both protocols are described below.

I. Discontinuous density Percoll gradients

Materials and reagents

- Anticoagulated blood (EDTA, heparin, citrate)
- Dextran T-500 (Amersham Pharmacia Biotech), 3% solution in 0.85% NaCl
- Ficoll-Paque or Histopaque, density 1.077 g/ml. (Amersham Pharmacia Biotech or Sigma)
- HBSS (Ca^{2+} and Mg^{2+} free) supplemented with 5% FBS (HBSS-5% FBS)
- Gelatin, 0.1% solution in HBSS (Ca^{2+} and Mg^{2+} free)
- 10× HBSS (Ca^{2+} and Mg^{2+} free)
- Percoll (Amersham Pharmacia Biotech). Mix Percoll with 10× HBSS to make a stock solution with a density of 1.124 g/ml (as determined by pycnometer).
- Diff-Quik or Wright-Giemsa stain (Fisher)
- Trypan blue stain
- Centrifuge
- 15- and 50-ml centrifuge tubes
- Graduated cylinder

- Pipettes
- 10-ml syringe and tubing

Protocol

1. Perform venipuncture and obtain approximately 180 ml of blood. Use 0.1% EDTA, heparin, or citrate as anticoagulant.
2. Prepare a 3% (w/v) solution of dextran in 0.85% NaCl. Mix equal parts of blood and dextran solution. Allow red blood cells (RBCs) to sediment for 45 min at room temperature using 50-ml centrifuge tubes or a graduated cylinder.
3. After a clean interface between the leukocyte-rich plasma and erythrocytes has formed, carefully aspirate the upper layer (plasma) to a centrifuge tube and pellet cells by centrifuging at 250 × *g* for 10 min at 4°C. Discard supernatant.
4. Resuspend the cell pellet in a volume of saline solution equal to the starting volume of blood.
5. Transfer approximately 40 ml of the cell suspension into a 50-ml conical centrifuge tube. Carefully underlayer with 10 ml of Ficoll-Paque solution, using a pipette or syringe with attached tubing. Centrifuge at 400 × *g* for 40 min at 20 °C.
6. Remove the mononuclear cell layer at the interface. Collect the granulocyte pellet and wash two times in HBSS-5% FBS solution. Count and resuspend to a density of 2×10^7 cells/ml in HBSS-5% FBS.
7. Prepare Percoll solutions of the following densities: 1.090, 1.095, and 1.100 g/ml. Dilute the stock Percoll with HBSS to create solutions of the desired densities.
8. Prepare a multiple discontinuous density gradient. Starting with the solution of the lowest density (1.090 g/ml), underlayer with the solutions of increasing density using flexible, 21-gauge tubing connected to a syringe. Use a 15-ml conical centrifuge tube.
9. Carefully layer the granulocyte suspension on top of the gradient.
10. Centrifuge at 700 × *g* for 20 min. Collect the cell band at the 1.095/1.100 g/ml interface and wash with a 0.1% gelatin solution in HBSS.
11. Lyse contaminating RBCs by resuspending the cell pellet in 4.5 ml of ice-cold water for 30 s. Immediately equilibrate the tonicity of the solution by addition of 0.5 ml of a 10× HBSS (Ca^{2+} and Mg^{2+} free).
12. Count and check viability using trypan blue stain.
13. The content of esoniphils can be determined in cytospins stained with Diff-Quik or Wright-Giemsa stain.

Comments

1. Ammonium chloride lysis has been reported to negatively influence eosinophil cytokine-mediated viability.[9]
2. Expect a purity of 90 to 95%.

II. Magnetic cell sorting with anti-CD16 beads

Materials and reagents

- Anticoagulated blood (EDTA, heparin, citrate)
- HBSS (Ca^{2+} and Mg^{2+} free)
- HBSS (Ca^{2+} and Mg^{2+} free) containing 2% FBS (HBSS-2% FBS)
- Gelatin, 0.1% solution in HBSS (Ca^{2+} and Mg^{2+} free)
- 10× HBSS (Ca^{2+} and Mg^{2+} free)
- Percoll, diluted in HBSS to 1.090 g/ml (Amersham Pharmacia Biotech)
- Anti-CD16 immunomagnetic beads (Miltenyi Biotec)
- Diff-Quik or Wright-Giemsa stain (Fisher)
- Trypan blue stain
- Centrifuge
- Magnetic separation apparatus (MACS™, Miltenyi Biotec)
- MACS column type C (Miltenyi Biotec)
- Platform rocker
- 50-ml conical centrifuge tubes
- Pipettes

Protocol

1. Perform venipuncture and obtain approximately 100 ml of blood.
2. Dilute blood with an equal volume of HBSS (Ca^{2+} and Mg^{2+} free).
3. In a 50-ml centrifuge tube, layer two parts of diluted blood (e.g., 20 ml) onto one part (e.g., 10 ml) of Percoll (1.090 g/ml). Centrifuge at 700 *g* for 20 min at room temperature to fractionate leukocytes.
4. Aspirate the band at the interface, which should contain the mononuclear cells. Collect the pellet and wash two times with HBSS (Ca^{2+} and Mg^{2+} free).
5. Lyse erythrocytes by resuspending the pellet in 27 ml of ice-cold water for 30 s. Immediately equilibrate tonicity with 3 ml of 10× HBSS (Ca^{2+} and Mg^{2+} free).
6. Wash cells two times in HBSS (Ca^{2+} and Mg^{2+} free) containing 2% FCS. Count cells and resuspend to approximately 2×10^8 cells/tube.
7. Centrifuge and resuspend cell pellet in 50 µl HBSS-2% FBS buffer. Add 50 µl of anti-CD16 immunomagnetic beads. Incubate for 40 min at 4°C on a platform rocker.
8. Prepare the magnetic column by rinsing with the following solutions: 70% ethanol, HBSS, and HBSS-2% FBS buffer. Store rinsed column in HBSS-2% FBS buffer at 4°C.
9. Add 4 ml of HBSS to the cell suspension and load onto the column. Run the cell suspension through the column while exposed to the magnetic field in order to remove the neutrophils, following the manufacturer's recommendations. Collect the flow-through.
10. Centrifuge cells. Count and assess viability and eosinophil purity by trypan blue and Diff-Quik or Wright-Giemsa staining, respectively.

Comment

Expect a yield of approximately 3 to 5×10^6 cells with a purity of >99%.

III. Functional assessment

A. Generation of LTC_4

Leukotriene C_4 (LTC_4) is an arachidonic acid-derived mediator produced by activated mast cells, basophils, and eosinophils. LTC_4 and its derived metabolites, LTD_4 and LTE_4, are powerful mediators of bronchoconstriction, and play a major role in the pathogenesis of asthma.[2,3] The generation of LTC_4 by eosinophils after stimulation with a variety of agents (e.g., IL-5, GM-CSF, Zymosan, A23187 ionophore) can be assessed in conjunction with commercially available RIA kits for LTC_4.[6]

Materials and reagents

- Eosinophil preparation (as described)
- LTC_4 RIA kit (Amersham Pharmacia; Du Pont NEN)
- PBS supplemented with 1 m*M* Ca^{2+}, 20 m*M* serine, and 5 m*M* glutathione
- Serum-treated Zymosan (STZ).[6] Resuspend Zymosan (Sigma Z-4250) to a concentration of 5 mg/ml in HBSS, boil for 10 min, and wash two times with HBSS. Opsonize the washed Zymosan particles by incubating in the presence of fresh human AB serum (0.5 ml serum per 2.5 g of Zymosan) for 30 min at 37°C and wash two times in cold HBSS. Resuspend in HBSS to 5 mg/ml and use immediately.
- A2318 Calcium ionophore (Sigma C-7522). Prepare a 1-mg/ml stock solution in DMSO and freeze at –80°C.
- Recombinant human IL-5 or GM-CSF (e.g., Pharmingen; R & D Systems)
- Water bath or incubator at 37°C
- Centrifuge

Protocol

1. Equilibrate isolated eosinophils to a cell density of 1×10^6/ml in warm (37°C) PBS supplemented with Ca^{2+}, serine, and glutathione.
2. Pre-incubate eosinophils at 37°C for 30 min in the presence or absence of cytokines (IL-5, GM-CSF [100 p*M*]).
3. Add STZ particles (0 to 0.5 mg/ml, final) or A23187 (0 to 25 μ*M*, final) and incubate cells for 20 min at 37°C.
4. Stop the activation by adding an equal volume of ice-cold reaction buffer (RIA kit). Centrifuge to remove cells and collect the supernatant.
5. Assay for LTC_4 according to procedure described in RIA kit.

B. Superoxide anion (O_2^-) generation

Similar to neutrophils, eosinophils can also generate O_2^- following activation.[5] Generation of O_2^- can be assayed by measuring the superoxide dismutase (SOD)-inhibitable reduction of ferricytochrome *c*, as described by Pick and Mizel.[10]

Materials and reagents

- Eosinophil preparation (as described)
- HBSS containing 0.1% gelatin
- Fe^{3+} Cytochrome *c*, 10 m*M* solution (e.g., horse heart, Type VI, Sigma C-7752)
- Cytochalasin B (Sigma, C-6762)
- Superoxide dismutase, 5 mg/ml (SOD; bovine erythrocytes, Sigma S-2515)
- Phorbol myristate acetate (PMA). Prepare a 5-mg/ml stock solution in DMSO and store at –80°C.
- A23187 Ionophore (Sigma). Prepare a 1-mg/ml stock solution in DMSO and freeze at –80°C.
- Formyl-methionyl-leucyl-phenylalanine (FMLP). Prepare a 10-m*M* stock solution in DMSO and freeze at –80°C.
- STZ (see above protocol)
- 96-well plates
- Multi-well spectrophotometer (550 nm)
- Incubator (37°C)
- Pipettes

Protocol

1. Plate eosinophils (1×10^5/well) in a final volume of 200 µl of HBSS containing 0.1% gelatin, 100 µ*M* cytochrome *c*, and 5 µg/ml cytochalasin B (unless PMA is used as the activator).
2. Initiate the reaction by addition of the desired stimulus. For example, PMA (0 to 10 ng/ml), A23187 (0 to 0.25 µ*M*), FMLP (0 to 10 µ*M*), or STZ (0 to 0.5 mg/ml). Reactions should be performed in duplicate and against identical control wells containing 20 µg/ml SOD.
3. Reactions should also include controls for spontaneous release (unstimulated cells).
4. Mix and immediately measure absorbance in a multi-well spectrophotometer at 550 nm.
5. Read absorbances every 10 min for 1 h, incubating the plate at 37°C between readings.
6. Calculate O_2^- generation using an extinction coefficient of 21.1×10^{-3} *M*/cm for reduced cytochrome *c*. Results are expressed as nmoles of cytochrome *c* reduced/5×10^5 cells/30 min, after subtraction of the SOD and spontaneous release controls.

C. Eosinophil peroxidase adhesion assay

Adhesion of eosinophils to collagen or to human vascular endothelial cells (HUVEC) is measured in 96-wells, as described by Sedgwick et al.[6] The detection of the bound eosinophils is based on the measurement of eosinophil peroxidase activity.[4]

Materials and reagents

- Eosinophil preparation (as described)
- HUVEC (ATCC) cultures
- Collagen (calf skin, type I, Sigma C-9791). 50 μg/ml solution in acidified water.
- HBSS and HBSS containing 0.1% gelatin
- Eosinophil activators (PMA, A23187, FMLP; see above protocol)
- Substrate solution: 1 m*M* H_2O_2, 1 m*M* *o*-phenylenediamine (Sigma P-1526), 0.1% Triton in 55 m*M* Tris buffer, pH 8.0
- 96-well tissue culture plates
- Incubator (37°C)
- Pipettes
- Multi-well spectrophotometer

Protocol

1. Prepare 96-well plates by coating with collagen or HUVEC monolayers. For collagen, coat wells by incubating with 30 μl/well of collagen solution for 30 min at 37°C. For HUVEC monolayers, culture HUVEC cells (2×10^5/ml) in collagen-coated wells for 24 to 48 h before the adherence assay.
2. Wash wells with warm (37°C) HBSS before assay.
3. Prepare an eosinophil suspension in HBSS containing 0.1% gelatin (1×10^5 eosinophils/ml). Plate 100 μl/well of the eosinophil suspension into the collagen- or HUVEC-coated wells, followed by 10 μl of activator (see above protocol). Incubate for 20 min at 37°C. Include unstimulated control wells.
4. Wash wells three times with warm (37°C) HBSS to remove nonadherent eosinophils.
5. Add 100 μl of HBSS-0.1% gelatin to each well. Add 100 μl of the original eosinophil suspension to empty wells as a control for total activity.
6. Prepare substrate solution (1 m*M* H_2O_2, 1 m*M* *o*-phenylenediamine, 0.1% Triton in 55 m*M* Tris buffer, pH 8.0). Add 100 μl per well.
7. Incubate for 30 min at room temperature and stop the reaction with 50 μl/well of a 3*M* H_2SO_4 solution.
8. Read absorbance in a multi-well spectrophotometer at 490 nm.

9. Calculate adherence as follows:% adherence = [(activated OD_{490} − spontaneous OD_{490})/total OD_{490} of 1×10^4 eosinophils] × 100. Alternatively, the number of eosinophils can be calculated from a standard curve prepared with known amounts (10^2 to 10^4) of eosinophils.

Comment

This assay does not measure neutrophil myeloperoxidase and does not result in eosinophil degranulation.[4]

References

1. Weller, P. E., Immunobiology of eosinophils, *N. Eng. J. Med.*, 324, 1110, 1991.
2. Corrigan, C. J. and Kay, A. B., T cells and eosinophils in the pathogenesis of asthma, *Immunol. Today*, 13, 501, 1992.
3. Frick, W. E., Sedgwick, J. E., and Busse, W. W., The appearance of hypodense eosinophils in antigen-dependent late phase asthma, *Am. Rev. Respir. Dis.*, 139, 1401, 1989.
4. Sedgwick, J. E., Calhoun, W. J., Vrtis, R. F., Bates, M. E., McAllister, P. K., and Busse, W. W., Comparison of airway and blood eosinophil function after *in vivo* antigen challenge, *J. Immunol.*, 149, 3710, 1992.
5. Sedgwick, J. E., Vrtis, R. F., Gourley, M. F., and Busse, W. W., Stimulus-dependent differences in superoxide anion generation by normal human eosinophils and neutrophils, *J. Allergy Clin. Immunol.*, 81, 876, 1988.
6. Sedgwick, J. B., Shikama, Y., Nagata, M., Brener, K., and Busse, W. W., Effect of isolation protocol on eosinophil function: Percoll gradients versus immunomagnetic beads, *J. Immunol. Reagents*, 198, 15, 1996.
7. Casale, T. B., Erger, R. A., and Rozell, M. D., Eosinophils isolated by magnetic cell sorting respond poorly to lipid chemoattractants, *Ann. Allergy Asthma Immunol.*, 83, 127, 1999.
8. Hansel, T. T., Pound, J. D., Pilling, D., Kitas, G. D., Salmon, M., Gentle, T. A., Lee, S. S., and Thompson, R. A., Purification of human blood eosinophils by negative selection using immunomagnetic beads, *J. Immunol. Reagents*, 122, 97, 1989.
9. Ide, M., Weiler, D., Kita, H., and Gleich, G. J., Ammonium chloride exposure inhibits cytokine-mediated eosinophil survival, *J. Immunol. Reagents*, 168, 187, 1994.
10. Pick, E. and Mizel, D., Rapid microassays for the measurement of superoxide and hydrogen peroxide production by macrophages in culture using an automatic enzyme immunoassay reader, *J. Immunol. Reagents*, 46, 211, 1981.

chapter four

Enrichment of human basophils

Basophils are the least abundant of the circulating leukocytes, and thus are relatively difficult to purify. Isolation protocols based on density-gradient centrifugation have been reported to result in basophil enrichments of up to 50%, although significant contamination with other cell types, particularly lymphocytes, still exists.[1] One key consideration during the enrichment of basophils is preventing their stimulation and degranulation. The method described is based on the report of Leonard et al.[2]

Materials and reagents

- Venous blood, freshly drawn
- 0.1 *M* EDTA solution, pH 7.4
- Percoll gradients. Prepare a Percoll (Amersham Pharmacia Biotech) stock solution by mixing 90 ml Percoll, 9 ml 10× HBSS, 1 ml 0.25 *M* HEPES buffer pH 7.3, and 0.4 ml 1*M* HCl. Dilute with 1× HBSS to prepare solutions of 1.070 g/ml (24 ml Percoll stock + 20 ml HBSS); 1.079 g/ml (27 ml Percoll stock + 15.9 ml HBSS); and 1.088 g/ml (23 ml Percoll stock + 10 ml HBSS).[1]
- HBSS (without Ca^{2+}, Mg^{2+}, and phenol red)
- 15- and 50-ml conical centrifuge tubes
- Refrigerated centrifuge
- Pipettes

Protocol

1. Prepare ten tubes with discontinuous Percoll gradients formed by sequentially adding to the bottom of the tube, with the aid of a 5-ml syringe and tubing, 4 ml of 1.070 g/ml, 4 ml of 1.079 g/ml, and 3 ml of 1.088 g/ml.
2. Obtain blood (40 ml) and immediately mix in a 50-ml tube with 1/10th the volume of 0.1 *M* EDTA solution, pH 7.4.

3. Layer 4 ml of the blood mixture over each of the 10 tubes containing the discontinuous Percoll gradients. Centrifuge at 300 × *g* for 25 min at room temperature.
4. Carefully remove tubes from the centrifuge. Remove each cell band with 1-ml disposable pipettes and pool similar bands using 15-ml centrifuge tubes. Most basophils should be present at the top of the 1.079 g/ml layer.
5. Wash cells two times with 10 ml of HBSS (centrifuge at 300 × *g* for 7 min at 4°C).
6. Stain aliquot of each preparation with Wright-Giemsa stain and determine basophil purity.

Comments

1. Use of heparin as an anticoagulant results in a lower percentage of basophils banding on top of the 1.079 g/ml layer.
2. Several techniques have been reported for the further purification of basophils after Percoll gradient centrifugation, including negative selection by panning with anti-T cell antibodies[3] or positive selection by the somewhat selective adherence of basophils to glass beads.[4]

References

1. Metcalfe, D. D., Isolation of human basophils, in *Current Protocols in Immunology*, Coligan, J.E., Kruisbeek, A. M., Margulies, D. H., Shevach, E. M., and Strober, W., Eds., Greene Publishing and Wiley Interscience, New York, 1991, 7.24.1.
2. Leonard, E. J., Roberts, R. L., and Skeel, A., Purification of human blood basophils by single step isopycnic banding on Percoll, *J. Leukocyte Biol.*, 35, 169, 1984.
3. Landry, F. J. and Findlay, S. R., Purification of human basophils by negative selection, *J. Immunol. Reagents*, 63, 329, 1983.
4. Pruzansky, J. J. and Patterson, R., Enrichment of human basophils, *J. Immunol. Reagents*, 44, 183, 1981.

chapter five

Macrophages

Contents

Macrophages have been considered to be the cells more or less responsible for the regulation of defense reactions since the pioneering work of Ellie Metchnikoff.[1] For more than a century, detailed microscopic analysis has suggested that free and fixed macrophages are involved in processes of tissue turnover, which include tissue remodeling during embryogenesis and metamorphosis, tissue destruction and repair subsequent to injury and infection, and tissue renewal, such as the removal of senescent or malignant cells. Reawakening of scientific interest in macrophages was brought about by Zanvil A. Cohn's studies.[2] To be sure, macrophages are not accessory cells only. Among their many special characteristics, one distinctive feature is responsible for the key role of macrophages in immune reactions: their presence in most, if not all, tissues of the living body. One can find them almost everywhere: in liver, spleen, bone marrow, lymph nodes, uterus, brain, thymus, bronchus- and gut-associated lymphoid tissue, and in many other organs and tissues. Their main function is to monitor and regulate circulating body fluids (both lymph and blood) and to react, both positively and negatively, to any changes (see Reference 3 for a review). As the first line of defense, they engulf and digest foreign particles, debris from dead cells, damaged erythrocytes, etc., both nonspecifically as well as by binding via their cell surface receptors for the Fc portion of immunoglobulin molecules and complement fragments.

The term macrophages corresponds to the final, mature stage of cells derived from pluripotent stem cells localized in bone marrow. The most immature cell type is the monoblast, which further divides into two promonocytes. These cells give rise to the monocyte, which resides in bone marrow for some period, during which it matures. Following the first steps in maturation, monocytes leave the bone marrow by a rather random process and via blood circulation they migrate into various tissues and organs of the body. When they enter the particular tissue, they are functionally less active than "older," resident macrophages. Readers seeking more information about development of cells of the myeloid lineage should review the excellent papers written by van Furth's group.[4] As macrophages develop under the instructive conditions of the local microenvironment, one can clearly understand that they differ from one another based on their origin. Therefore, macrophages also are commonly and rightfully named according to their final localization sites, i.e., peritoneal macrophages, osteoclasts in bones, microglial cells of the brain, etc.

The question whether macrophages localized in various tissues are at the final stage of their differentiation or if they are a partially self-renewing population is still not completely resolved. According to one hypothesis, the population of peritoneal macrophages is continuously replenished by blood monocytes entering the peritoneal cavity; a second hypothesis supports the idea that the resident peritoneal macrophages form a self-sustaining population.

To make the whole situation even more complex and more complicated, there is always the question of normal (i.e., macrophages under steady-state conditions) vs. activated (stimulated, elucidated, exudated, etc.) macrophages. The yield of the easily accessible and most commonly used peritoneal macrophages is approximately 2×10^6 per mouse, which is insufficient for large-scale experiments. Therefore, scientists often use various sterile stimulants to boost both the yield and the percentage of macrophages. The most common stimulant is probably the thioglycollate medium, which can substantially increase the yield (up to 3×10^7 depending on mouse strain). For more details about various kinds of activating stimuli and the properties of differently potentiated macrophages, see References 5 to 7. Activated macrophages differ from their normal counterparts in all respects; they are incredibly versatile cells possessing a bristling armamentarium. They are significantly larger, attach faster to the substrate, and have a more active metabolism. The major reason for these differences is that macrophages differentiate in a stepwise fashion. Therefore, their development reflects multiple signals, such as cytokines, bacterial products, etc. Some agents, e.g., thioglycollate medium or proteose peptone, stimulate the macrophages into an intermediate and noncytolytic state. Some factors (such as LPS) are potent enough to transform the macrophages directly into the activated state or can later trigger the next developmental step and result in a fully tumoricidal state.[8] One must keep in mind that a wide range of drugs and substances are known to substantially affect macrophage activities. This list would be extremely broad: muramyl peptides, lipopolysaccharide, interferon, concanavalin A, indomethacin, etc. Therefore, any substance (and quite often even a simple experimental manipulation) used during isolation might influence their properties.[9]

For years, phagocytosis was proposed as the main activity and *raison d'etre.* However, macrophages are much more involved in both the afferent and efferent branches of immune response. Other principal macrophage functions involve pinocytosis, degradation of ingested material, chemotaxis, antimicrobial activity, secretion of immune and other factors (more than 100 substances are produced and released by macrophages), antigen processing and presentation, cooperation with T and B lymphocytes, and cell lytic activity.

I. Isolation of macrophages

The most commonly used macrophages in immunological studies are mouse macrophages isolated from the peritoneal cavity. The simple technique of their isolation is given in Chapter 1. Macrophages can be isolated from many different organs, including bone marrow, spleen, or peripheral blood (in the form of monocytes). Due to the rather low percentage of monocytes in most of the organs and tissues, the isolation process also involves a further purification step. Readers interested in purification of macrophages from heterologous cell suspension should read *Reagents in Cellular Immunology.*[10]

A. Isolation of bone marrow macrophages

Materials and reagents

- Mice
- RPMI 1640 medium supplemented with antibiotics, 10% FCS and glutamine
- Phosphate-buffered saline (PBS)
- Mouse rIL-3 (Genzyme)
- Squeeze bottle with 70% ethanol
- Scissors and forceps

- Cutting board or paper towels
- 60 × 15-mm petri dishes
- 3-ml syringes
- 22G1 needles
- 15-ml conical centrifuge tube
- 25- and 75-cm^2 tissue culture flasks
- Dispase grade II, 1.0 mg/ml (Boehringer-Mannheim), sterilize and warm to 37°C
- Rubber policeman
- Humidified 37°C, 5% CO_2 incubator

Protocol

(*Note:* All materials and reagents must be sterile and proper aseptic technique must be used when handling the cells.)

1. Isolate bone marrow cells as described in Chapter 1.
2. Resuspend cells in supplemented RPMI medium to a final concentration of 5×10^6 cell/ml.
3. Cultivate 2×10^7 cells in 25-cm^2 tissue-culture flasks containing 10 ml supplemented RPMI medium with 10 ng/ml IL-3 for 24 h.
4. Transfer nonadherent cells to 75-cm^2 tissue-culture flasks containing 10 ml supplemented RPMI medium with 10 ng/ml IL-3 and cultivate for an additional 4 days.
5. Add an additional 10 ml of supplemented RPMI media with 10 ng/ml IL-3 and cultivate for another 3 days.
6. Remove culture supernatant and wash adherent cells with 20 ml PBS.
7. Add 5 ml of diluted dispase into each flask and incubate 5 min at 37°C.
8. Gently shake the flasks and remove the cells by scraping using sterile rubber policeman.
9. Add 15 ml of supplemented RPMI media to each flask and harvest the cells.
10. Centrifuge the cells at $1000 \times g$ at 4°C for 10 min.
11. Resuspend in 5 ml of desired medium and count.

Comments

1. For tissue culture studies, perform the entire isolation in a sterile hood and use only sterile instruments, medium, and dishes.
2. Keep cells on ice.
3. Colony-stimulating factor-1 (CSF-1) can be used instead of IL-3.[11–13]

II. Activation of macrophages

As mentioned in the introduction to this chapter, macrophages significantly differ in their properties and functions, based on their developmental stage.[5,14] For some experiments (especially elimination of bacterial, parasitic, or tumor targets), we might need activated macrophages instead of normal cells. Activation of macrophages, either *in vitro* or *in vivo*, might be advantageous. *In vivo* activation of peritoneal macrophages using a thioglycollate medium is described in Chapter 1.

Materials and reagents[15]

- Peritoneal exudate cells isolated 3 days after injection of thioglycollate medium (Chapter 13)
- 24-well flat-bottom tissue culture wells
- Dispase grade II, 1.0 mg/ml (Boehringer-Mannheim), sterilize and warm to 37°C
- DMEM medium with 10% FCS and antibiotics
- 15-ml conical centrifuge tubes
- LPS
- Murine recombinant IFN-γ
- Rubber policeman
- Humidified 37°C, 5% CO_2 incubator

Protocol

(*Note:* All materials and reagents must be sterile and proper aseptic technique must be used when handling the cells.)

1. Centrifuge the cells in DMEM medium and concentrate to 1×10^6 cells/ml.
2. Add 500 µl cells into each well and incubate in a humidified 37°C, 5% CO_2 incubator for 60 min.
3. Wash away the nonadherent cells by repeated washes with 1.5 ml of DMEM medium.
4. Add 2 µl of 500 U/ml IFN-γ into each well and incubate in a humidified 37°C, 5% CO_2 incubator for 4 h.
5. Repeat Step 3.
6. Add 1 µl of 5 µg/ml LPS into each well and incubate in a humidified 37°C, 5% CO_2 incubator for 60 min.
7. Repeat Step 3.
8. Add 0.5 ml of diluted dispase into each well and incubate 5 min at 37°C.
9. Gently shake the plates and remove the cells by scraping, using sterile rubber policeman.
10. Add 2 ml of supplemented DMEM medium to each well and harvest the cells.

Comments

1. Resulting macrophages are fully activated both for cytotoxicity and for production of reactive nitrogen. If only cytotoxic activity is desired, substitute addition of 5 µl of 500 U/ml IFN-γ instead of LPS.
2. Do not use mouse strains with defective LPS genes such as C3H/HeJ.

A. Leucine aminopeptidase activity[16,17]

Leucine aminopeptidase is a marker of macrophage activation and maturation.

Materials and reagents

- Macrophages (mouse, human, or rat)
- PBS
- Leucine-4-methoxy-2-naphtylamine (Sigma)
- *N*,N-dimethylformamide (Kodak Chemical)
- 5-Nitrosalicylaldehyde (Kodak Chemical)
- 0.15 *M* 2-[N-morpholino]ethanesulfonic acid, pH 6.5
- Sodium acetate buffer (see Chapter 13)
- Sodium acetate buffer with 0.02% Triton X-100
- 12 × 75-mm polystyrene test tubes
- Water bath
- Cell cytofluorometer

Protocol

1. Isolate macrophages as described in Chapter 1.
2. Dilute cells to a concentration 2×10^6/ml in PBS.
3. Dissolve 1 mg leucine-4-methoxy-2-naphtylamine in 20 µl *N*,N-dimethylformamide.
4. Dissolve 5-nitrosalicylaldehyde in 20 µl *N*,N-dimethylformamide.
5. Add 5 ml of 0.15*M* 2-[N-morpholino]ethanesulfonic acid, pH 6.5, to each of the above solutions.
6. Mix the two solutions and add 1 ml of the resulting mixture to 1 ml cells.
7. Incubate for 20 min in a water bath at 37°C.
8. Wash the cells by centrifugation in 0.1*M* sodium acetate buffer with 0.02% Triton X-100, pH 5.2, at 300 × *g* for 5 min at 4°C.
9. Resuspend the cells in 200 µl of cold 0.1*M* sodium acetate buffer.
10. Evaluate the cells in flow cytofluorometer using 352 to 363 excitation and 525 emission band pass filter.

III. Production of oxidative metabolism intermediates

It is well established that macrophages produce various oxidative metabolism intermediates during phagocytosis and cytotoxic reaction. Both reactions have distinct biochemical pathways[18,19] and are independently regulated.[20] Readers seeking more information about the formation of nitrogen intermediates should consult the excellent review by Green and Nacy.[21]

The ligand-receptor interactions result in a substantial outburst of metabolic energy. During initial stages of phagocytosis, the cells exhibit a large increase in oxygen consumption, hexosemonophosphate shunt activity, and production of active oxygen species.[22-24] The process is known as the respiratory or oxidative burst. The generation of superoxide radicals is catalyzed by a membrane-localized NADPH-oxidase, which is triggered by an appropriate membrane stimulation. This oxidase transfers electrons from cytosolic NADPH to extracellular oxygen, producing H_2O_2. This is necessary for the killing of invading microorganisms, but at the same time causes inflammation and tissue damage. The simultaneous presence of several microbicidal possibilities is advantageous in that a wide range of microorganisms could be killed. The multiplicity of the attack system is most probably the result of a long-lasting host–parasite coevolution.[25]

The oxidative burst is not necessarily dependent on phagocytosis, and thus constitutes an important characterization of phagocyte activity. Production of O_2^- (superoxide anion) is the initial step in the conversion of oxygen to hydrogen peroxide and hydroxyl radicals, which are potent microbicidal metabolites. Therefore, the level of superoxide anion production is a valuable indicator for antimicrobial potential of macrophages. The most reliable technique for measurement of O_2^- activity is the measurement of superoxide dismutase-inhibitable reduction of Fe^{3+} cytochrome *c* to Fe^{2+}. For more information about the metabolism of reactive oxygen metabolites, see References 19 and 26 to 28.

A. Nitrite formation[30]

Materials and reagents

- Mouse peritoneal cells (Chapter 1)
- DMEM medium with 5% FCS and antibiotics
- Murine recombinant IFN-γ
- 1 m*M* N-monomethyl-L-arginine acetate (Calbiochem) in sterile PBS
- LPS
- 2 m*M* $NaNO_2$ (Sigma) in DMEM medium
- Greiss reagent solutions (Chapter 13)
- 96-well flat-bottom tissue culture plates
- Humidified 37°C, 5% CO_2 incubator
- Microtiter plate reader

Protocol

(*Note:* All materials and reagents must be sterile and proper aseptic technique must be used when handling the cells.)

1. Centrifuge the cells in DMEM medium and concentrate to 1×10^6 cells/ml.
2. Add 0.1 ml of peritoneal cells into wells of the 96-well plate and incubate at 37°C for 2 h.
3. Wash away the nonadherent cells by repeated washes with 0.2 ml of DMEM medium.

4. To appropriate wells add the following:
   N-monomethyl-L-arginine acetate to 50 to 250 μ*M* final concentration
   IFN-γ to 0.1 to 10 U/ml final concentration
   LPS to 2 to 10 ng/ml final concentration
5. Incubate in a humidified 37°C, 5% CO_2 incubator for 48 h.
6. Collect 50 μl of culture supernatants from each well and transfer to wells of a new 96-well flat-bottom tissue culture plate.
7. Prepare serial dilutions of $NaNO_2$ solution from 125 to 1 μ*M* final concentration.
8. Add 50 μl of each Greiss reagent solution to wells with samples and the $NaNO_2$ dilutions.
9. Measure absorbance at 550 nm using a microtiter plate reader. Use absorbance of the $NaNO_2$ dilutions as a standard curve.

Comments

1. Tumor necrosis factor can be used instead of LPS.
2. Five different sets of cells should be used: peritoneal cells alone; cells treated with IFN-γ; cells treated with LPS; cells treated with IFN-γ and LPS; and cells treated with N-monomethyl-L-arginine acetate, IFN-γ, and LPS.
3. If testing for cytotoxic activity, the appropriate target cells should be added in Step 4.
4. For optimal results, add Greiss reagent solution before $NaNO_2$ solution in Step 8.
5. Prepare samples in triplicate.
6. Do not use mouse strains with defective LPS genes such as C3H/HeJ.
7. Macrophages previously activated *in vivo* or isolated from other organs or tissues might respond differently to the activating agents.

B. *Superoxide anion*[29]

Materials and reagents

- Mouse peritoneal cells (Chapter 1)
- RPMI 1640 medium with 5% FCS and antibiotics
- Hanks' balanced salt solution (HBSS), with Ca^{2+} and Mg^{2+}, phenol red-free
- Fe^{3+} Cytochrome *c* (Chapter 13)
- Superoxide dismutase (Sigma), 1 mg/ml in distilled water
- Murine recombinant IFN-γ
- Phorbole 12-myristate 13-acetate (PMA, Sigma), 2 mg/ml in DMSO
- Sodium dithionite (Sigma), 1 mg/ml in HBSS, with Ca^{2+} and Mg^{2+}, phenol red-free
- 96-well flat-bottom tissue culture plates
- Humidified 37°C, 5% CO_2 incubator
- Microtiter plate reader

Protocol

(*Note:* All materials and reagents must be sterile and proper aseptic technique must be used when handling the cells.)

1. Centrifuge the cells in RPMI 1640 medium and concentrate to 1×10^6 cells/ml.
2. Add 0.1 ml of cell suspension into wells of a 96-well flat-bottom tissue culture plate.
3. Add INF-γ to appropriate wells at final concentration 0.1 to 10 U/ml and incubate in a humidified 37°C, 5% CO_2 incubator for 48 h.
4. Remove medium and wash the cells three times with HBSS.
5. To one set of samples, add 75 μl HBSS and 25 μl Fe^{3+} cytochrome *c* solution.
6. To a second set of samples, add 50 μl HBSS, 25 μl Fe^{3+} cytochrome *c* solution and 25 μl superoxide dismutase.
7. Do not forget to set cell-free wells from Steps 5 and 6.
8. Add 10 μl PMA into each well and incubate at 37°C for 60 min.
9. Transfer 150 μl of each supernatant to wells of a new 96-well flat-bottom microtiter plate.
10. Measure absorbance at 550 nm using a microtiter plate reader.
11. Add 50 μl sodium dithionite into each well.
12. Repeat Step 10.
13. Calculate the amount of released O_2^-[28] as follows:
    (a) Determine total Fe^{3+} cytochrome *c* reduction by showing reading from Step 10 as a percentage of reading from Step 12 (A).
    (b) Subtract the values for the superoxide dismutase-treated samples from the corresponding experimental values (B).
    (c) Calculate:
    nmol $O_2^- = A \times B \times 10.9$

Comments

1. PMA can be stored at –80°C for at least 1 year.
2. The total amount of released nmol O_2^- can be calculated per number of macrophages, per min of incubation, or per mg of cell protein.
3. Macrophages previously activated *in vivo* or isolated from other organs or tissues might respond differently to the activating agents.

IV. Phagocytosis

The term phagocytosis comes from the Greek *phagein*, meaning "to eat," and is used to describe the uptake of solid particular material such as bacteria, erythrocytes, viruses, fungi, and other organic and inorganic materials. Although the ingestion of foreign material by cells was observed and mentioned by several earlier scientists (e.g., Koch and Roser described cells filled with anthrax bacilli, but the significance eluded them[31]) it was first described by Metchnikoff more than 100 years ago.[1] The elegant demonstrations of rose thorn introduced under the skin of starfish larvae formed the basis of the phagocyte theory.

A series of distinct events is associated with the internalization and subsequent processing of particulate material by phagocytes. Each phase of phagocytosis is a complex cellular process having special functional features and different metabolic requirements, which calls for the close cooperation of different extra- and intracellular factors and consequently displays varying degrees of sensitivity to microenvironmental factors. Readers seeking more details regarding these sequences should see Reference 28. Phagocytosis of particles represents one of the several functional parameters used for characterization of macrophages. Phagocytosis, the original function of which in unicellular and primitive *Metazoa* was feeding of cells, acquired a new meaning during evolution. As a process for removal of bacteria, damaged cells, and other foreign particles, it became an important mechanism of nonspecific immunity. By removing bacteria, parasites, degradation products of metabolism, and other foreign material, it became an indispensable part of immune reactions. It is probably one of the most widely occurring cellular functions. Although most probably all eukaryotic cells demonstrate this "primitive" function, it is especially important for macrophages, leukocytes, endothelial and epithelial cells, oocyte, and yolk sac cells.

To achieve more objective evaluation of this process, the choice of the particles is extremely important. Apart from various strains of bacteria and erythrocytes, particles commonly classified as inert are employed as reagents for determination of phagocytic activity. This group of prey includes silica, carboxylated metals, $CdCO_3$ microcrystals, and latex microspheres. However, certain types of cells avidly form nonspecific bonds with such particles depending on their surface charge and hydrophobic character, exhibiting no ability for subsequent ingestion.[14,32]

A. Phagocytosis of bacteria[33]

Phagocytosis of bacteria belongs to the classical pool of immunological techniques dating well back into the 19th century. Almost all types of bacteria can be used for this method. The most commonly used evaluation is examination of cells under a light microscope, but a fluorescent microscope, a fluorocytometer, or an electron microscope are also used.

Materials and reagents

- Macrophages (mouse or human; see Chapter 1)
- Bacterial culture
- Normal serum (mouse or human according to the macrophages used)
- Phosphate-buffered saline (PBS) with 5% FCS
- Balanced salt solution (BSS)

- 30% Sucrose in PBS
- Diff-Quik (Baxter)
- 12 × 75-mm polypropylene tubes
- Microscope slides and cover glass
- End-over-end shaker
- Cytospin cytocentrifuge
- Refrigerated centrifuge
- Parafilm foil

Protocol

1. Centrifuge the cells two times in BSS at 250 × *g* for 10 min at 4°C.
2. Adjust the concentration at a 2.5×10^7 cell/ml.
3. Add 100 µl of macrophage suspension to each tube.
4. Dilute the bacterial suspension 1:10 in BSS.
5. Add 100 µl of bacterial suspension into each tube.
6. Add 50 µl of fresh serum and 750 µl of BSS.
7. Seal the tubes with Parafilm foil.
8. Place the tubes on laboratory shaker and rotate end-over-end for 20 min at 37°C.
9. Wash the macrophages three times in BSS, as shown in Step 1. Resuspend the cells in 1 ml of cold BSS and underlay with 1 ml of 30% sucrose.
10. Centrifuge for 8 min at 250 × *g* at 4°C. Remove both sucrose and BSS and resuspend the pellet in 2 ml of PBS with 5% FCS.
11. Cytocentrifuge the cells using Cytospin cytocentrifuge, according to the manufacturer's specifications.
12. Stain the slides with Diff-Quik solution, according to the manufacturer's specifications.
13. Evaluate the phagocytosis under the microscope. Calculate the results either as % of phagocytosing cells or as phagocytic index.

Comments

1. To be sure that the bacteria are used as a single cell suspension, it is beneficial to vortex the bacterial suspension vigorously.
2. Do not incubate for more than 30 min;[33] the killing of bacteria by macrophages is quite rapid.
3. Additional staining, such as Giemsa-Romanowsky or Wright stain, can also be used.
4. Phagocytic index can be calculated as follows:

phagocytic index (PI) = % of macrophages with at least one bacterium

× mean number of bacteria per positive macrophage

5. Use C3-sufficient serum from the same species as the macrophages. It might be better to use a fresh serum. However, the use of an aliquot from a large batch of frozen serum prepared and tested previously is also advantageous.
6. It is important not to use antibiotics during the entire experiment. The effects of small amounts of antibiotics on the ability of macrophages to ingest and kill bacteria have been described.[34]

B. Phagocytosis of FITC-labeled material[33]

To distinguish properly between only adherent and fully engulfed prey might be quite difficult and often requires an experienced scientist. Various techniques have been suggested in the past, including extensive washing procedures, destroying the binding of attached particles, use of particles with zero nonspecific binding,[6] or quenching of outside particles. In studies of phagocytosis, fluorescence techniques have been used for differentiation of intracellular and extracellular particles. This experimental approach is often based on the observation that bacteria or fungi vitally stained with acridine orange and examined in ultraviolet or blue excitation light will show green (if viable) or red (if dead) color.

Another possible approach is the utilization of fluorescent synthetic microbeads and fluorescence quenching effect. Using this technique, one can easily differentiate between the attached and the ingested particles in the individual phagocytes. The principle of this technique is based on the observation that some dyes (e.g., crystal violet) quench the fluorescence of free and/or cell-bound fluoresceinated particles, whereas the internalized particles remain fluorescent. The phenomenon called excitation-energy transfer is responsible for this quenching effect.[35] The excitation energy of the fluorochrome molecule is transferred to a molecule of the quenching agent attached to the same polymer molecule. The transfer occurs when the absorption spectrum of the quenching agent overlaps the emission spectrum of fluorescent label. This technique was originally described by Hed for a test for phagocytosis based on FITC-conjugated bacteria or yeast as the prey and crystal violet as the quenching agent.[36] Later, some modifications to this method were published. Loike and Silverstein used trypan blue for quenching of glutaraldehyde-fixed erythrocytes.[37] The advantage of trypan blue is its active exclusion from the viable cells; ingested erythrocytes do not change their fluorescence during the entire time the cell is alive.

We found that the simple double staining of bacteria works best. In this technique, extracellular FITC-labeled bacteria (green) are counterstained with ethidium bromide and thus acquire a red color. Fully internalized bacteria remain green, as ethidium bromide does not penetrate the cell membrane.

Materials and reagents

- Macrophages (mouse or human; see Chapter 1)

- Heat-killed bacterial culture (most commonly used *Listeria monocytogenes* can be obtained from ATCC; strain 15313)
- Normal serum (mouse or human, according to the macrophages used)
- Fluorescein isothiocyanate isomer 1 (Sigma), 0.1 mg/ml in 0.1 *M* $NaHCO_3$, pH 9.0
- Phosphate-buffered saline (PBS) with 5% FCS
- PBS with 5% FCS and 5 m*M* glucose
- Ethidium bromide (Sigma)
- Balanced salt solution (BSS)
- 30% sucrose in PBS
- 12 × 75-mm polypropylene tubes
- Microscope slides and cover glass
- End-over-end shaker
- Cytospin cytocentrifuge
- Refrigerated centrifuge
- Parafilm foil
- Fluorescence microscope

Protocol

1. Wash heat-killed bacteria in PBS by centrifugation for 3 min at 12,000 × *g* at room temperature. Discard the supernatant.
2. Resuspend the pellet in 1 ml of FITC in $NaHCO_3$.
3. Incubate 60 min at room temperature.
4. Wash labeled bacteria five times by centrifugation for 3 min at 12,000 × *g* at room temperature. Discard the supernatant and resuspend in PBS.
5. Perform phagocytosis reaction up to Step 9.
6. Centrifuge for 8 min at 250 × *g* at 4°C. Remove both sucrose and BSS and resuspend the pellet in 1 ml of PBS with 5% FCS and 5 m*M* glucose.
7. Remove 100 µl aliquots and mix them with ethidium bromide to 50 µg/ml final concentration.
8. Place 1 drop on a microscope slide and cover with coverslip.
9. Evaluate under fluorescence microscope.

Comments

1. Ethidium bromide is a potent teratogen; use proper precautions.
2. Ethidium bromide crosses the cell membrane very slowly; therefore, there is about a 120-min interval for evaluation.
3. The same technique can be used for evaluation of phagocytosis of yeasts.
4. It is important not to use antibiotics during the entire experiment. The effects of small amounts of antibiotics on the ability of macrophages to ingest and kill bacteria have been described.[32]

C. Phagocytosis of protamine–heparin aggregates[38]

Protamine–heparin aggregates are minute particles of microscopic size, produced *both in vivo* and *in vitro* by the rapid interaction of polycationic protamine and polyanionic heparin. They quickly penetrate the glomerular ultra filtration barrier and accumulate in the capillary wall.[39] These complexes exhibit some of the characteristics of antigen–antibody complexes and are successfully used as an experimental model of immune-complex disease.[40] The complexes produced *in vitro* assume a spherical or slightly oval shape with a polyanion surface.

Materials and reagents

- Macrophages (mouse or human; see Chapter 1)
- Protamine sulfate (ICN Biomedicals)
- Heparin (Sigma)
- PBS
- RPMI 1640 medium supplemented with 5% FCS
- Refrigerated centrifuge
- 15-ml centrifuge tubes

Protocol

1. Wash the cells two times in RPMI 1640 medium by centrifugation at $500 \times g$ for 10 min at 4°C.
2. Dilute cells in RPMI 1640 to a concentration 2×10^6/ml.
3. Add 1.5 ml of protamine–heparin suspension (7.5 mg protamine sulfate and 750 IU heparin) to the cells and incubate 60 min at 37°C.
4. Wash the cells three times in RPMI 1640 medium by centrifugation at $500 \times g$ for 10 min at 4°C. Use the pellet for immunocytochemistry.

Comments

1. This technique is particularly suitable for evaluation of phagocytosis by means of electron microscopy. Unfortunately, to describe all necessary techniques of electron microscopy is outside the scope of this book. Readers seeking more information should consult Reference 41.
2. In addition to electron microscopy, the use of immunocytochemistry is also recommended.

D. Cytometric assessment of phagocytosis[42]

Materials and reagents

- Fresh whole blood in heparin
- Ice bath

- Phagotest kit (Orpegen, Heidelberg, Germany) containing opsonized FITC-labeled *Escherichia coli*, quenching solution, DNA staining solution, lysing solution, and washing solution
- Vortex
- 5-ml polystyrene tubes
- Water bath with shaker
- Pipettes
- Centrifuge
- Flow cytometer

Protocol

1. Use only fresh blood with heparin.
2. Vortex blood and place 100 µl aliquots in test tubes.
3. Put blood, bacteria, quenching solution, and washing solution in an ice bath for 20 min.
4. Dilute the bacteria to 10^9/ml in washing solution and vortex well.
5. Add 20 µl of bacteria to each tube with blood.
6. Vortex the tubes.
7. Incubate the tubes for 10 min at 37°C in the shaking bath.
8. Put the tubes into an ice bath, add 100 µl quenching solution, and vortex.
9. Add 3 ml washing solution into each tube and centrifuge at 250 × *g* for 5 min at 4°C.
10. Discard the supernatant and repeat Step 9.
11. Add 3 ml of lysing solution to each tube, vortex, and incubate for 20 min at room temperature.
12. Centrifuge at 250 × *g* for 5 min at 4°C.
13. Repeat Step 9.
14. Add 100 µl DNA staining solution to each tube.
15. Vortex and incubate for 10 min, protected from light in an ice bath.
16. Measure on flow cytometer using 488 nm excitation and 530-nm and 630-nm filters.

Comments

1. Do not use EDTA or citrate, as they negatively influence phagocytosis.
2. Use all solutions ice cold.
3. After Step 13, samples can be stored overnight.

V. Pinocytosis

Pinocytosis was originally discovered by Haeckel[43] in *Tethys fibria,* but not until 1930 was the microscopy of single cells advanced sufficiently to allow Lewis[44] to describe the uptake of extracellular fluid in full detail. The name pinocytosis, from the Greek *pinos*, meaning "to drink," describes the uptake of anything from droplets of fluid, coloids, or immunocomplexes to soluble macromolecules. As a rule, pinocytic vesicle is always filled with fluid. Two major types of pinocytosis have been described:

1. Receptor-mediated pinocytosis
2. Fluid-phase pinocytosis

The differences between these two types are simple: an uptake by means of fluid-phase pinocytosis is nonsaturable and cannot concentrate the material; an uptake via receptor-mediated pinocytosis is saturable and able to concentrate the ligand intracellularly.[45]

A. *Pinocytosis assay*[16]

Materials and reagents

- Cells (neutrophils, monocytes, macrophages)
- Fluorescein-dextran (Sigma)
- PBS gel buffer, pH 7.4 (see Chapter 13)
- 4% Paraformaldehyde
- PBS with 0.1% bovine serum albumin
- Stimuli (see *Comment*)
- Ice
- Eppendorf microcentrifuge tubes
- Microcentrifuge
- Pipettes
- Water bath

Protocol

1. Wash cells twice in PBS gel buffer by centrifuging at 250 × g for 10 min at 4°C.
2. Resuspend cells at 2×10^6/ml in PBS gel buffer.
3. Place the cells on ice.
4. Prepare serial dilutions of stimulus in PBS gel buffer.
5. Prepare 1:5 dilution of dextran-FITC in PBS gel buffer.
6. Prewarm 4% paraformaldehyde and cell suspension at 37°C for 10 min.
7. To the appropriate Eppendorf tube, add either 50 μl of stimulus or PBS gel and 50 μl of diluted dextran-FITC.
8. Mix well and incubate at 37°C for 5 min.
9. Add 400 μl of cell suspension to each tube.
10. Incubate for 10 min at 37°C.
11. Add 500 μl of 4% paraformaldehyde to each tube.
12. Incubate at room temperature for 30 min.
13. Wash cells three times in PBS with bovine serum albumin by centrifuging at 250 × *g* for 10 min at 4°C.

14. Resuspend the cells in 500 μl PBS with bovine serum albumin.
15. Evaluate on flow cytometer.

Comment

Various stimulating agents such as PMA, cytochalasin B, or FMLP can be used.

VI. Killing of bacteria[46]

Materials and reagents

- Macrophages (mouse or human; see Chapter 1)
- Live bacterial culture
- Balanced salt solution (BSS)
- Normal serum (mouse or human, according to the macrophages used)
- Cold BSS with 10% normal serum
- Tryptic soy agar culture plates
- 2-ml conical tubes with cap
- 30% sucrose in PBS
- 10 × 75-mm polypropylene tubes
- 100-mm glass tubes with caps (Corning)
- End-over-end shaker
- Pipettes
- Ice

Protocol

(*Note:* All materials and reagents must be sterile and proper aseptic technique must be used when handling the cells.)

1. Dilute bacterial culture overnight 1:300 in BSS.
2. Dilute macrophages to 2×10^7/ml in BSS.
3. To a 10 × 75-mm tubes, add 100 μl macrophages, 300 μl diluted bacteria, 50 μl cold normal serum, and 550 μl BSS.
4. Close the tubes with Parafilm foil.
5. Place the tubes on laboratory shaker and rotate end-over-end for 20 min at 37°C.
6. Wash the macrophages three times in BSS by centrifuging at 250 × *g* for 10 min at 4°C.
7. Resuspend the cells in 1 ml of cold BSS and underlay with 1 ml of 30% sucrose.
8. Centrifuge for 8 min at 250 × *g* at 4°C. Remove both sucrose and BSS and resuspend the pellet in 2 ml of BSS with 10% normal serum.
9. For each experimental tube, prepare four glass tubes with 900 μl sterile distilled water each. Starting with 100 μl original cell mixture, make 1:10 serial dilutions into these glass tubes.

10. Plate 100 µl (in duplicate) on rewarmed (37°C) tryptic soy agar culture plates.
11. Seal the tubes with undiluted samples and incubate 120 min at 37°C.
12. Put all tubes on ice.
13. Prepare diluted samples as described in Step 9, and plate them as described in Step 10.
14. Invert the plates and incubate for 24 h at 37°C.
15. Count the colonies on every plate.
16. Compare the number of colonies from Step 13 with background values from Step 10.

Comments

1. Various different bacteria can be used; the most commonly used bacteria are *Escherichia coli*, *L. monocytogenes*, and *Staphylococcus*.
2. Use C3-sufficient serum from the same species as the macrophages. It might be better to use a fresh serum. However, the use of an aliquot from a large batch of frozen serum prepared and tested previously is also advantageous.
3. To be sure that the bacteria are used as a single cell suspension, it is beneficial to vortex the bacterial suspension vigorously.
4. Set up the tubes in triplicate.
5. Use a new pipette for each serial dilution.
6. It is important not to use antibiotics during the entire experiment. The effects of small amounts of antibiotics on the ability of macrophages to ingest and kill bacteria have been described.[32]

VII. Killing of bacteria — colorimetric evaluation[47]

The principle of this technique is the measurement of living bacteria by colorimetric evaluation of their ability to reduce MTT to formazan.

Materials and reagents

- Macrophages (mouse or human; see Chapter 1)
- Live bacterial culture (diluted to 10^7/ml)
- Normal serum (mouse or human, according to the macrophages used)
- MTT (Sigma); 5 mg/ml in PBS
- RPMI 1640 medium without phenol red, supplemented with 5% normal serum
- 2.95% (w/v) tryptose phosphate broth (Difco)
- 5% saponin (Sigma)
- Flat-bottom 96-well microtiter plates
- 15-ml conical centrifuge tubes
- Pipettes
- Humidified 37°C 5% CO_2 incubator
- ELISA plate reader

Protocol

1. Wash the cells two times in RPMI 1640 medium by centrifugation at 300 × *g* for 10 min at 4°C.
2. Dilute macrophages to 1×10^6/ml in RPMI 1640.
3. Add 100 µl of cells/well of the 96-well plate. Prepare two identical plates.
4. Add 10 µl of diluted bacteria into each well and incubate both plates for 20 min at 37°C in humidified 37°C 5% CO_2 incubator.
5. Centrifuge both plates at 250 × *g* for 10 min at 4°C.
6. Remove all supernatant.
7. Add 100 µl RPMI 1640 to each well including four blank (empty) wells. Repeat Step 5.
8. Add 20 µl of 5% saponin into all wells of plate A.
9. Incubate plate A for 60 s at room temperature.
10. Add 100 µl tryptose phosphate broth to each well and store at 4°C.
11. Incubate the second plate (B) for 120 min at 37°C in humidified 37°C 5% CO_2 incubator.
12. Incubate both plates for an additional 4 h at 37°C
13. Add 15 µl of diluted MTT into each well and incubate for 20 min at 37°C in humidified 37°C 5% CO_2 incubator.
14. Measure the absorbance using an ELISA reader at 570 nm.
15. Prepare a standard curve by incubating known numbers of bacteria with MTT and measuring the formazan formation using an ELISA reader at 570 nm. Compare the numbers of bacteria at the beginning (plate A) and end (plate B) of the incubation by extrapolation from the prepared standard curve.

Comments

1. Store dissolved MTT in dark container for 2 months at 4°C.
2. MTT is a carcinogen. Use all appropriate precautions.
3. Design the experiment in triplicate.
4. It is important not to use antibiotics during the entire experiment. The effects of small amounts of antibiotics on the ability of macrophages to ingest and kill bacteria have been described.[34]

VIII. Microbicidal assay[46]

Materials and reagents

- Isolated peripheral blood monocytes (see Chapter 1)
- *Staphylococcus aureus* (ATCC 27217)
- Mueller-Hinton Broth (Difco)
- 0.1% Gelatin in distilled water

- HBSS
- 0.05% Gentian violet in 2% acetic acid (Sigma)
- Blood agar plates (Difco)
- 12 × 75-mm polystyrene tubes
- Centrifuge
- Incubator with rotation
- Control sera pooled from at least five normal donors, aliquoted and stored at –70°C
- Pipettes
- Spectrophotometer
- Vortex

Protocol

(*Note:* All materials and reagents must be sterile and proper aseptic technique must be used when handling the cells. In addition, all material and reagents must be LPS-free.)

1. One day before the experiment, inoculate 10 ml of Mueller-Hinton broth with 5 µl *S. aureus* and incubate overnight.
2. Aliquot 2.5 ml of 0.1% gelatin into tubes.
3. Isolate monocytes as described in Chapter 1.
4. Centrifuge the broth containing *S. aureus* at 900 × *g* for 15 min.
5. Resuspend the pellet in 10 ml of 0.1% gelatin and centrifuge again at 900 × *g* for 15 min.
6. Resuspend the bacteria in 1 ml of 0.1% gelatin.
7. Prepare 1:30, 1:60, 1:120, and 1:240 dilutions of bacteria in 0.1% gelatin.
8. Measure the dilutions using a spectrophotometer at 620 nm. Use 0.1% gelatin as a blank. 1:120 dilution generally corresponds to 10^8 bacteria per ml and should give about 0.1 OD. If not, prepare the dilution that will give 0.1 OD.
9. Thaw an aliquot of serum and maintain at 4°C.
10. Label blood agar plates in duplicates (0, 30, 60, 120, and 240) for both bacteria only (D) and bacteria with monocytes (E).
11. Dilute the monocytes to 2.94×10^6/ml with HBSS.
12. Prepare polystyrene test tubes as follows:
    A — 850 µl of HBSS and 150 µl of serum
    B — 850 µl of monocytes and 150 µl of serum
13. Add 25 µl of *S. aureus* dilution (with OD 0.1 at 620 nm) to each tube and vortex.
14. Remove 10 µl from each tube and add to the tube containing 2.5 ml of 0.1% gelatin (labeled the same way as blood agar plates).
15. Place the tubes in a 37°C incubator with rotation and incubate for 30 min.
16. Vortex the tubes labeled D0 and E0, transfer 25 µl in duplicate to appropriately labeled blood agar plates, and spread the inoculum across the plate.

17. Remove the tubes from the incubator after the end of incubation.
18. Repeat Step 16 with tubes labeled D30 and E30.
19. At 30-min intervals, repeat Step 16 with tubes labeled D60 and E60, and D90 and E90, respectively.
20. Incubate the plates at 37°C overnight.
21. Count the numbers of colonies on each plate.

Comment

Peripheral blood neutrophils can be used instead of blood monocytes.

IX. Antiviral activity of macrophages[48]

The major antiviral activity of macrophages is exerted via production of interferon. However, the situation with demonstrated antiviral activity without detectable levels of interferon has been repeatedly described.[49] The assay given below utilizes resistance of macrophages to infection with vesicular stomatitis virus.

Materials and reagents

- Murine macrophages
- RPMI 1640 medium supplemented with 10% FCS, 2 m*M* glutamine, 30 m*M* HEPES, and antibiotics
- Vesicular stomatitis virus
- PBS
- Crystal violet (0.1% in 0.1 *M* acetic acid)
- Methanol
- Aspirator
- 5% Formaldehyde
- 96-well flat-bottom tissue culture plates
- Plastic squirt bottle
- Pipettes
- Humidified 37°C 5% CO_2 incubator
- ELISA plate reader

Protocol

(*Note:* All materials and reagents must be sterile and proper aseptic technique must be used when handling the cells. In addition, all material and reagents must be LPS-free.)

1. Wash the cells two times in RPMI 1640 medium by centrifugation at $300 \times g$ for 10 min at 4°C.
2. Dilute macrophages to 1×10^6/ml in RPMI 1640.
3. Add 200 µl of cells into each well of a 96-well tissue culture plate and incubate for 4 h at 37°C.

4. Dilute vesicular stomatitis virus in RPMI 1640 medium to 2×10^5/ml.
5. Aspirate and discard medium from each well and add 100 µl of viral suspension to each well.
6. Incubate for 24 h at 37°C.
7. Wash each well with 200 µl PBS.
8. Aspirate and discard PBS from each well and add 200 µl of 5% formaldehyde to each well.
9. Incubate 10 min at room temperature.
10. Aspirate and discard formaldehyde and add 50 µl crystal violet to each well.
11. Incubate 10 min at room temperature.
12. Aspirate and discard crystal violet. Wash the plate six times with tap water using plastic squirt bottle.
13. Add 100 µl 100% methanol to each well.
14. Read absorbance at 595 nm using an ELISA plate reader.

Comments

1. Vesicular stomatitis virus is a potentially dangerous pathogen. Use all appropriate precautions.
2. Uninfected control cells will have the highest absorbance.
3. C3H/HeJ mice-derived peritoneal macrophages are more susceptible to vesicular stomatitis virus infection.[50]
4. It is important to be sure that macrophages have formed an intact monolayer. Check all wells microscopically after Step 3. If necessary, incubate for a longer period of time.

References

1. Metchnikoff, E. E., Researches on the ultracellular digestion of invertebrates, *J. Microscop. Sci.*, 24, 155, 1884.
2. Cohn, Z. A., The fate of bacteria within the phagocytic cells. II. The modification of intracellular degradation, *J. Exp. Med.*, 117, 43, 1963.
3. Fornusek L. and Větvička, V., *Immune System Accessory Cells*, CRC Press, Boca Raton, 1992.
4. van Furth, R., Identification of mononuclear phagocytes: overview and definitions, in *Reagents for Studying Mononuclear Phagocytes*, Adams, D. O., Edelson, P. J., and Koren, H. S., Eds., Academic Press, New York, 1981, 243.
5. Větvička, V., Zeleny, V., and Viklicky, V., Cell surface changes of the mouse peritoneal macrophages after proteose peptone or thioglycollate induction, *Fol. Biol.*, 25, 156, 1979.
6. Fornusek, L., Větvička, V., and Kopecek, J., Differences in phagocytic activity of methacrylate copolymer particles in normal and stimulated macrophages, *Experientia*, 37, 418, 1981.
7. Morahan, P. S., Volkman, A., Melnikoff, M. J., and Dempsey, W. L., Macrophage heterogeneity, in *Macrophages and Cancer*, Hoppner G. H. and Fulton, A., Eds., CRC Press, Boca Raton, 1988, 1.

8. Hibbs, J. B., Taintor, R. R., Chapman, H. A., and Weinberg, J. B., Macrophage tumor killing: influence of the local environment, *Science*, 197, 279 1977.
9. Haskill, S., Johnson, C., Eierman, D., Becker, S., and Warren, K., Adherence induces selective mRNA expression of monocyte mediators and proto-oncogenes, *J. Immunol.*, 140, 1690, 1988.
10. Fernandez-Botran, R. and Větvička, V., *Reagents in Cellular Immunology*, CRC Press, Boca Raton, 1995.
11. Moore, R. N., Larson, H. S., Horohov, S. W., and Rouse, B.T., Endogenous regulation of macrophage proliferative expansion by colony stimulating factor-induced interferon, *Science*, 223, 178, 1984.
12. Warren, M. K. and Vogel, S. N., Bone marrow-derived macrophages: development and regulation of differentiation markers by colony stimulating factor and interferon, *J. Immunol.*, 134, 982, 1985.
13. Fortier, A. H. and Falk, L. A., Isolation of murine macrophages, in *Current Protocols in Immunology*, Coligan, J. E., Kruisbeek, A. M., Margulies, D. H., Shevach, E. M., and Strober, W., Eds., John Wiley & Sons, New York, 1998, 14.1.1.
14. Větvička, V., Fornusek, L., Kopecek, J., and Prikrylova, D., Phagocytosis of 2-hydroxyethylmethacrylate copolymer particles by different types of macrophages, *Folia. Biol.*, 29, 424, 1983.
15. Fortier, A. H., Activation of murine macrophages, In *Current Protocols in Immunology*, Coligan, J. E., Kruisbeek, A. M., Margulies, D. H., Shevach, E. M., and Strober, W., Eds., John Wiley & Sons, New York, 1998, 14.4.1.
16. Robinson, J. P., *Handbook of Flow Cytometry Reagents*, Wiley-Liss, New York, 1993.
17. Mahoney, K. H., Miller, B. E., and Heppner, G. H., FACS quantization of leucine aminopeptidase and acid phosphatase on tumor-associated macrophages from metastatic and nonmetastatic mouse mammary tumors, *J. Leukocyte Biol.*, 38, 573, 1985.
18. Archer, S., Measurement of nitric oxide in biological models, *FASEB J.*, 7, 349, 1993.
19. Nathan, C. F., Release of hydrogen peroxide, in *Reagents for Studying Mononuclear Phagocytes*, Adams, D. O., Edelson, P. J., and Koren, H., Eds., Academic Press, New York, 1981, 498.
20. Nathan, C. F. and Gabay, J., Antimicrobial mechanisms of macrophages, in *Mononuclear Phagocytes*, van Furth, R., Ed., Kluwer Academic Publishers, Amsterdam, 1992, 259.
21. Green, S. J. and Nacy, C. A., Antimicrobial and immunopathologic effects of cytosine-induced nitric oxide synthesis, *Curr. Opin. Infect. Dis.*, 6, 384, 1993.
22. Badwey, J. A. and Karnovsky, M. L., Active oxygen species and the function of phagocytic leukocytes, *Ann. Rev. Biochem.*, 49, 695, 1980.
23. Karnovsky, M. L., Metabolic basis of phagocytic activity, *Physiol. Rev.*, 42, 1422, 1962.
24. Michl, J., Pieczonka, M. M., Unkeless, J. C., and Silverstein, S. C., Effects of immobilized immune complexes on Fc- and complement-receptor function in resident and thioglycollate-elicited mouse peritoneal macrophages, *J. Exp. Med.*, 150, 607, 1979.
25. Větvička, V. and Sima, P., *Evolutionary Mechanisms of Defense Reactions*, Birkhauser Publishers, Basel, 1998.
26. Klebanoff, S. J., Oxygen metabolites from phagocytes, in *Inflammation: Basic Principles and Clinical Correlates*, Gallin, J. I., Goldstein, I. M., and Synderman, R., Eds., Raven Press, New York, 1992, 541.

27. Rotrosen, D., The respiratory burst oxidase, in *Inflammation: Basic Principles and Clinical Correlates,* Gallin, J. I., Goldstein, I. M., and Synderman, R., Eds., Raven Press, New York, 1992, 589.
28. Větvička, V. and Fornusek, L., Limitations of transmembrane transport in drug delivery, *Crit. Rev. Ther. Drug Carrier Syst.,* 5, 141, 1988.
29. Green, S. J., Aniagolu, J., and Raney, J. J., Oxidative metabolism of murine macrophages, in *Current Protocols in Immunology,* Coligan, J. E., Kruisbeek, A. M., Margulies, D. H., Shevach, E. M., and Strober, W., Eds., John Wiley & Sons, New York, 1998, 14.5.1.
30. Pick, E. and Mizel, D., Rapid microassays for the measurement of superoxide and hydrogen peroxide production by macrophages in culture using an automatic enzyme immunoassay reader, *J. Immunol. Meth.,* 46, 211, 1981.
31. Hirsch, J. G., Host resistance to infections diseases — a centennial, in *Advances in Host Defence Mechanisms,* Gallin, J. I. and Fauci, A.S., Eds., Raven Press, New York, 1982, 1.
32. van Oss, C. J., Phagocytosis as a surface phenomenon, *Annu. Rev. Microbiol.,* 32, 19, 1978.
33. Campbell, P. A., Canono, B. P., and Drevets, D. A., Measurement of bacterial ingestion and killing by macrophages, in *Current Protocols in Immunology,* Coligan, J. E., Kruisbeek, A. M., Margulies, D. H., Shevach, E. M., and Strober, W., Eds., John Wiley & Sons, New York, 1998, 14.6.1.
34. Drevets, D. A., Canono, B. P., Leenen, P. J. M., and Campbell, P. A., Gentamicin kills intracellular *Listeria monocytogenes, Infect. Immun.,* 62, 2222, 1994.
35. Förster, T., Transfer mechanism of electronic excitation energy, *Radiat. Res.,* Supp. 2, 326, 1960.
36. Hed, J., The extinction of fluorescence by crystal violet and its use to differentiate between the attached and ingested microorganisms in phagocytosis, *FEMS Lett.,* 1, 357, 1977.
37. Loike, J. D. and Silverstein, S. C., A fluorescence quenching technique using trypan blue to differentiate between attached and ingested glutaraldehyde-fixed red blood cells in phagocytosing murine macrophages, *J. Immunol. Meth.,* 57, 373, 1983.
38. Rossmann, P. and Větvička, V., Phagocytosis of protamine–heparin aggregates by mouse peritoneal exudate cells, *Folia. Microbiol.,* 30, 319, 1985.
39. Sharon, Z., Schwartz, M. M., and Lewis, E. J., The glomerular localization and transport of aggregated protamine–heparin complexes, *Lab. Invest.,* 37, 43, 1977.
40. Seiler, M. W., Hoyer, J. R., and Krueger, T. E., Altered localization of protamine–heparin complexes in aminonucleoside nephrosis, *Lab. Invest.,* 43, 9, 1980.
41. Ogawa, K. and Barka, T., *Electron Microscopic Cytochemistry and Immunocytochemistry in Biomedicine,* CRC Press, Boca Raton, 1993.
42. O'Gorman, M.R., Phagocytic cell functions. II. Adhesion and phagocytosis, in *Manual of Clinical Laboratory Immunology,* Rose, N. R., de Macario, E. C., Folds, J. D., Lane, H. C., and Nakamura, R. M., Eds., American Society of Microbiology, Washington, D. C., 1997, 329.
43. Haeckel, E., *Die Radiolarien (Rhizopoda Radiaria),* G. Reiner, Berlin, 1862.
44. Lewis, W. H., Pinocytosis, *Bull. Johns Hopkins Hosp.,* 49, 17, 1931.
45. Steinman, R. M., Silver, J. M., and Cohn, Z. A., Pinocytosis in fibroblasts, quantitative studies *in vitro, J. Cell. Biol.,* 63, 949, 1974.

46. Campbell, D. E. and Douglas, S. D., Phagocytic cell functions. I. Oxidation and Chemotaxis, in *Manual of Clinical Laboratory Immunology,* Rose, N. R., de Macario, E. C., Folds, J. D., Lane, H. C., and Nakamura, R. M., Eds., American Society of Microbiology, Washington, D. C., 1997, 320.
47. Peck, R., A one-plate assay for macrophage bactericidal activity, *J. Immunol. Meth.,* 82, 131, 1985.
48. Falk, L. A., Measurement of interferon-mediated antiviral activity of macrophages, in *Current Protocols in Immunology,* Coligan, J. E., Kruisbeek, A. M., Margulies, D. H., Shevach, E. M., and Strober, W., Eds., John Wiley & Sons, New York, 1998, 14.9.1.
49. Vogel, S. N., Havell, E. A. and Spitalny, G. L., Monoclonal antibody-mediated inhibition of interferon γ-induced macrophage antiviral resistance and surface antigen expression, *J. Immunol.,* 136, 2917, 1986.
50. Vogel, S. N. and Fertsch, D., Macrophages from endotoxin-hyporesponsive (LPS^d) C3H/HeJ mice are permissive for vesicular stomatitis virus because of reduced levels of endogenous interferon: possible mechanism for natural resistance to infection, *J. Virol.,* 61, 812, 1987.

chapter six

B lymphocyte cloning

Contents

Growth and differentiation of hematopoietic cells are extremely complex processes involving interaction of numerous cell types and their products. The *in vivo* assays are crucial for our understanding of B lymphocyte developmental pathways. B lymphocytes are formed continuously throughout life. In mammals, during embryonic life, B lymphocytes are made in both the spleen and liver[1,2] within discrete microenvironments inside these organs. In adult mammals, cells with distinctive pre-B cell characteristics can be found in the bone marrow, but not in the spleen. Immunologists agree that bone marrow is the primary site of B cell production *de novo*. Eight different cell types (B lymphocytes, T lymphocytes, neutrophils, macrophages, eosinophils, erythrocytes, megakaryocytes, and mast cells) are made simultaneously in a rather complex and crowded environment, all of them originating from the same multipotent stem cells.

Stem cells are considered to be quiescent and self-renewing. The next developmental step includes committed progenitors, i.e., cells that are already committed into a lymphoid or myeolid lineage. Various intermediate cell types follow the developmental stages of hematopoietic development. In the case of B cell development these intermediate cell types are called early B lineage precursor, large pre-B precursor, small pre-B cell, newly formed B cell, and finally B lymphocyte.[3,4] The development of clonal cell culture systems capable of supporting the growth and development of lympho-hemopoietic progenitors provides an important tool for direct analysis of the early stages of the development of lymphocytes (see Figure 1).

I. Murine B lymphocyte cloning assay[5]

Approximately 3% of the sIg positive B lymphocytes divide to form colonies under these conditions. Surface Ig negative cells do not proliferate in semi-solid agar cultures. Detailed observations[5–7] show that a wide variety of B cell subpopulations, with the exception of pre-B cells, can be detected by this assay. An advantage of this simple experimental design is that we can directly study the effects of humoral factors such as antibodies or cytokines on individual cells.

Materials and reagents

- McCoy's medium supplemented with 15% fetal calf serum, sodium bicarbonate, sodium pyruvate, MEM essential amino acids, MEM non-essential amino acids, 2-mercaptoenthanol, L-glutamine, L-asparagine, and L-serine (see Chapter 13)
- Agar (Difco)
- Cells
- Erlenmeyer flasks
- 35-mm tissue culture petri dishes
- 15-ml tubes
- 5-ml pipettes
- LPS from *Salmonella typhosa* WO 901 (Difco)
- Water bath
- Incubator tray
- Humidified CO_2 incubator
- Stereoscopic dissecting microscope

Protocol

(*Note:* All materials and reagents must be sterile and proper aseptic technique must be used when handling the cells.)

1. Prepare complete McCoy's medium.
2. Melt agar (10× stock solution) in H_2O, 1.5 g/50 ml. Cool the agar to 40°C and mix 1:10 with complete McCoy's medium warmed to 37°C.

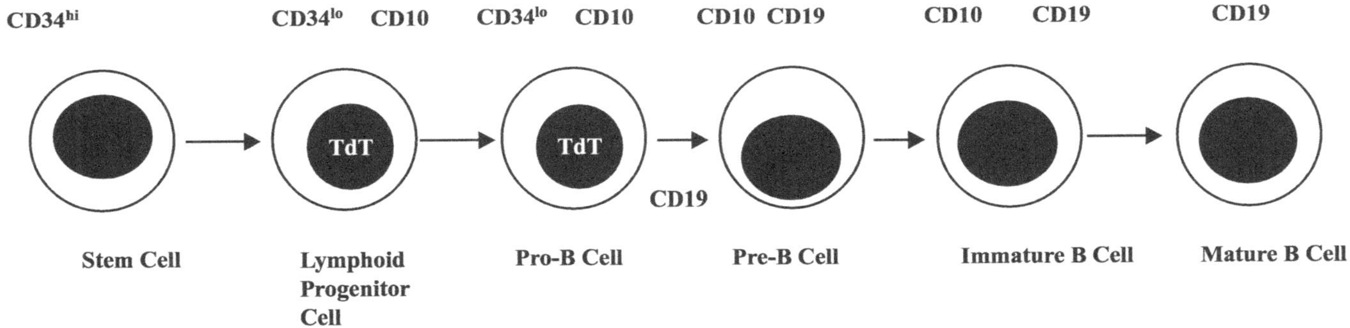

Figure 1 Some of the major stages of human B lymphocyte development. Loss or acquisition of certain membrane markers can be detected with monoclonal antibodies.

3. Isolate bone marrow cells (see Chapter 1) and prepare cell suspension in McCoy's medium at concentration 5×10^5 cells/ml.
4. Mix 0.1 ml of cell suspension with LPS (25 μg/plate) and 0.9 ml of agar solution.
5. Quickly pipette the warm solution of cells and medium into the dishes with a 5-ml pipette. Swirl the dishes to achieve thorough mixing.
6. Allow the dishes to stand undisturbed for 15 min at room temperature. Transfer the plates into CO_2 incubator and incubate at 37°C and 7% CO_2 for 6 days.
7. Count the number of colonies.

Comments

1. Clonal proliferation in semisolid agar cultures is dependent on water-soluble mitogens present in laboratory-grade agar.[5] However, if nothing else is added, colony size and numbers are suboptimal and the assay is nonlinear.
2. LPS potentiates colony formation by B cells from normal mice, but would not be expected to influence cultures of LPS-nonresponsive C3H/HeJ or C57BL/1 B cells.[8–10]
3. Sheep red blood cells (1% solution) can be used to overcome this problem.[8] Nevertheless, the B cells that are detected under these different conditions may not be completely overlapping.
4. If sheep erythrocytes are used instead of LPS, they have to be lysed by the addition of 0.4 ml of 3% glacial acetic acid in H_2O prior scoring.
5. Individual lots of FCS differ substantially in their ability to support B lymphocyte formation. It is therefore necessary to screen several batches of FCS from numerous suppliers and then reserve the adequate amount of the optimal FCS.
6. This technique works only for murine cells.

II. Proliferation of pre-B cell colonies in methylcellulose

Materials and reagents

- MEM medium supplemented with 100 μg/ml streptomycin and 100 U/ml penicillin
- Mouse IL-7
- FCS
- Methylcellulose
- 2-Mercaptoethanol
- 35 × 10-mm petri dishes
- 3-ml syringes
- 18G needles
- Humidified CO_2 incubator

Protocol

(*Note:* All materials and reagents must be sterile and proper aseptic technique must be used when handling the cells.)

1. Prepare MEM medium with 2.1% methylcellulose solution. Aliquot into 100-ml lots and store at –30°C.
2. Prepare MEM medium with 15% FCS, 5×10^{-5} *M* 2-mercaptoethanol, IL-7, and 0.8% methylcellulose (use 40 ml of 2.1% stock medium per 100 ml of final medium).
3. Isolate bone marrow cells (see Chapter 1) and prepare cell suspension.
4. Mix 3×10^5 of cells/plate in 0.1 ml with 2.9 ml of the 0.8% methylcellulose medium.
5. Using a 3-ml syringe and an 18-gauge needle, add 1 ml of cell suspension into two plates.
6. Incubate in CO_2 incubator at 37°C and 5% CO_2 for 6 days.
7. Count the number of colonies.

Comments

1. It is necessary to determine the optimal concentration of IL-7.
2. A wide range of cell concentration ranging from 10^3 to 10^6 is recommended.

III. Clonal assay for murine lymphohematopoietic progenitors[11]

The existence of pluripotent stem cells has long been postulated, based on either indirect observations or on *in vivo* experiments such as retroviral labeling of individual stem cells.[12] Two techniques for *in vitro* quantification of pluripotent stem cells have been recently developed: the first one involves a two-step methylcellulose clonal culture system;[13] the second one requires a co-culture of fetal progenitors with murine stromal cells.[14] A combination of two cytokines, stem cell factor, and one of interleukins IL-6, IL-11, or IL-12, supports proliferation and differentiation of the lymphohematopoietic progenitors. Subsequent micromanipulation of single cells in a two-step culture demonstrates the development of both lymphoid and myeloid cells from single lymphohematopoietic progenitors *in vitro.*

Materials and reagents

- Minimum essential medium (MEM) supplemented with 100 µg/ml streptomycin and 100 U/ml penicillin
- FCS, heat inactivated
- Bone marrow cells
- 2-Mercaptoethanol

- Bovine serum albumin, fraction V (BSA)
- Methylcellulose (1500 centipoise)
- 5-Fluorouracil (Sigma)
- 35-ml tissue culture dishes
- Stem cell factor (Sigma)
- Mouse IL-6 (100 ng/ml)
- Humidified CO_2 incubator
- 15-ml polystyrene tubes

Protocol

(*Note:* All materials and reagents must be sterile and proper aseptic technique must be used when handling the cells.)

1. Prepare 10% stock solution of BSA and 3% solution of methylcellulose[15] (see Chapter 13).
2. Harvest bone marrow cells from femurs and tibiae of mice 48 h after intravenous injection of 150 mg/kg 5-fluorouracil.[16] Isolate and subsequently purify the bone marrow cells (see Chapter 1).[17]
3. Prepare MEM medium supplemented with 25% FCS, 1% BSA, 1×10^{-4} *M* 2-mercaptoethanol, and 1.2% (w/v) methylcellulose.
4. Place 1.25 ml FCS, 0.5 ml BSA, 0.05 ml 1×10^{-2} *M* 2-mercaptoethanol, 2 ml 3% methylcellulose, IL-6, and 250 cells into 15-ml polystyrene tubes and adjust the volume to 5 ml with MEM. Shake well.
5. Pipette 1 ml of suspension into four dishes. Gently tilt the dishes to allow good distribution of mixture.
6. Incubate in CO_2 incubator at 37°C and 5% CO_2 for 8 to 13 days.

Comments

1. Individual lots of FCS differ substantially in their abilities to support colony formation. It is therefore necessary to screen several batches of FCS from numerous suppliers and then reserve the adequate amount of the optimal FCS.
2. The total number of bone marrow cells is significantly decreased by 5-fluorouracil treatment;[16] therefore, more mice should be used.
3. IL-11 (100 ng/ml) or IL-12 (10 ng/ml) can be used instead of IL-6.
4. These so-called primary colonies consist of committed B-lymphoid progenitors. It is possible to lift the individual colonies from the medium using a micropipette, wash them, and plate them again in secondary methylcellulose cultures (5% suspension with the addition of IL-7 and stem cell factor). These secondary colonies are formed by pre-B cells. Upon adoptive transfer into SCID mice, they reconstitute the production of serum immunoglobulins and spleen B cells.[13] Small colonies of myeloid cells can be easily distinguished.
5. Dissolve the 5-fluorouracil in sterile PBS at desired concentration by intensive shaking under warm running water.

IV. Clonal assay for fetal cells[18]

Bi-potent progenitor cells isolated from murine fetal tissue (liver, yolk sac) are used for this assay.[19] The stimulating factors are IL-7 in combination with stromal cell line S17. Later it was found that these conditions also are adequate for support of clonal growth of B cell progenitors from multipotent progenitors.[20] The presence of a stromal cell line can be replaced by adding IL-11 and mast cell growth factor.[21]

Materials and reagents

- OptiMEM medium (Gibco) supplemented with 100 μg/ml streptomycin and 100 U/ml penicillin
- 5×10^{-5} *M* 2-mercaptoethanol
- FCS
- S17 stromal cell line[22]
- Murine IL-7 (250 U/ml)
- LPS (10 μg/ml)
- γ-Irradiator
- 96-well tissue culture plates
- Humidified CO_2 incubator

Protocol

(*Note:* All materials and reagents must be sterile and proper aseptic technique must be used when handling the cells.)

1. Cells from 12-day-old mouse embryos are isolated from fetal liver by panning or magnetic beads separation on the basis of AA4.1 and Ly6A expression.
2. Incubate 1×10^3 S17 stromal cells/well at 37°C and 5% CO_2 for 12 h.
3. Irradiate the plates at 2000 rad.
4. Add isolated fetal liver cells in OptiMEM supplemented with 10% FCS and 250 U/ml IL-7. A minimum of three cell concentrations, i.e., 3, 10, and 30 cells/well, is recommended.
5. Incubate in CO_2 incubator at 37°C and 5% CO_2 for 10 days. On days 4 and 7, remove 100 μl of supernatant and replace with fresh OptiMEM medium containing IL-7.
6. Resuspend the cells from each well in 200 μl of OptiMEM medium, and transfer 25 μl into a new plate containing 1×10^3 irradiated S17 stromal cells/well in 200 μl medium supplemented with 10 μg/ml LPS.
7. Incubate in CO_2 incubator at 37°C and 5% CO_2 for 15 days.

Comments

1. Mitogen-responsive B cell progenitors can be identified as large blast colonies.

2. S17 stromal cells can be replaced by the addition of IL-11 (100 ng/ml) and MGF (100 ng/ml). When you use these factors, OptiMEM medium supplemented with IL-7, IL-11, and MGF must be used in Step 5.
3. A danger of overcrowding with myeloid progenitors can occur at higher cell concentrations. However, it is possible to transfer 25% of nonadherent cells into a new well.

V. Long-term bone marrow cultures[23]

For further studies of mechanisms by which cellular cooperation within the bone marrow microenvironment regulates development of B cells, it was necessary to find techniques allowing long-term cultures of bone marrow cells. Culture conditions first were described in 1977 by Dexter,[24] followed later by the technique established by Whitlock and Witte.[25] As both techniques use stromal cells as a feeder cell population, these reagents help to establish the stromal cell–B precursors relationship taking place in hematopoietic microenvironment. As was later established, stromal cells involve endothelial cells, reticular cells, and macrophages.[26] Each of these two techniques is unique in the final type of differentiated cells. Cells present in Dexter-type cultures include neutrophils, macrophages, and myeloid cell precursors; in Whitlock–Witte cultures only B lineage cells can be found. The possibility of establishing switch cultures transferring Dexter culture cells into Whitlock–Witte conditions gives the investigators another tool for studies of immature precursor cells development.[27]

A. Dexter myeloid cultures

Materials and reagents

- MEM medium supplemented with 100 μg/ml streptomycin and 100 U/ml penicillin
- Horse serum
- Hydrocortisone sodium succinate
- 25-cm^2 tissue culture flasks
- 3-ml syringes
- 25G needles
- Humidified CO_2 incubator

Protocol

(*Note:* All materials and reagents must be sterile and proper aseptic technique must be used when handling the cells.)

1. Prepare complete MEM medium supplemented with 20% horse serum, 10^{-6} *M* hydrocortisone sodium succinate, and antibiotics.
2. Add 8 ml of medium to 25-cm^2 tissue culture flasks.
3. Isolate bone marrow cells (see Chapter 1).

4. Add cells isolated from one femur into each flask. Total volume of medium should be 10 to 11 ml.
5. Incubate in CO_2 incubator at 33°C and 5% CO_2 for 7 days.
6. Remove half of the medium and nonadherent cells. Add 4 ml of fresh medium.
7. Incubate in CO_2 incubator at 33°C and 5% CO_2 for an additional 7 days.
8. Isolate fresh bone marrow cells as described earlier.
9. Remove all medium and add 10^6 of bone marrow cells into each flask. Total volume of medium should be 8 ml.
10. Repeat Step 7 every week. Nonadherent cells can be used in functional assay.

Comments

1. Dexter cultures can be maintained for several weeks.
2. Individual lots of horse serum differ substantially in their ability to support long-term bone marrow colonies. It is therefore necessary to screen several batches of horse serum from numerous suppliers and then reserve an adequate amount of the optimal horse serum.

B. Whitlock–Witte lymphoid cultures

Materials and reagents

- RPMI 1640 medium supplemented with 100 μg/ml streptomycin and 100 U/ml penicillin
- FCS
- 2-Mercaptoethanol
- 25-cm^2 tissue culture flasks

Protocol

(*Note:* All materials and reagents must be sterile and proper aseptic technique must be used when handling the cells.)

1. Prepare complete RPMI 1640 medium supplemented with 5% FCS, 5×10^{-5} *M* 2- mercaptoethanol, and antibiotics.
2. Isolate fresh bone marrow cells (see Chapter 1) and resuspend them at 1×10^6/ml in complete RPMI 1640 medium.
3. Add 13.5 ml of the cell suspension into each flask.
4. Incubate in CO_2 incubator at 37°C and 5% CO_2 for 3 days.
5. Add 5 ml of fresh medium and incubate in CO_2 incubator at 37°C and 5% CO_2 for 4 days.
6. Remove 14 ml of medium and add 10 ml of fresh medium.
7. Incubate in CO_2 incubator at 37°C and 5% CO_2.
8. Repeat Steps 6 (remove and add 7 ml of medium) and 7 twice a week.

Comment

Long-term cultures can be maintained for several months.

C. Establishment of human bone marrow stromal cells[28]

In order to successfully perform the final step from the mountain of data obtained on murine models to study development of human B lymphocytes, it was necessary to prepare techniques of cultivation of human bone marrow cells *in vitro*. The culture system described below requires the presence of stromal cells supporting the IL-7-dependent growth of human B cell precursors.[29,30]

Materials and reagents

- Ex-Cell 300 medium (JRH Bioscience) supplemented with 50 μg/ml streptomycin and 50 U/ml penicillin
- FCS
- Ficoll (Sigma)
- Hypaque-76 (Winthrop Pharmaceuticals)
- 75-cm^2 tissue culture flasks
- 50-ml polypropylene tubes
- Pasteur pipettes
- 0.5% w/v trypsin-EDTA solution
- Humidified CO_2 incubator

Protocol

(*Note:* All materials and reagents must be sterile and proper aseptic technique must be used when handling the cells.)

1. Prepare Ficoll-Hypaque solution $\varrho = 1.09$ (see Chapter 1).
2. Adult bone marrow is usually obtained as 10-ml aspirates in medium containing heparin.
3. Resuspend each ml of bone marrow cells in 2 ml of complete Ex-Cell medium supplemented with 2% FCS. Layer 30 ml of bone marrow cells in Ex-Cell medium over 15 ml of Ficoll-Hypaque solution in 50-ml polypropylene tubes.
4. Centrifuge at $500 \times g$ for 30 min.
5. Collect the low-density interface fraction of cells with a Pasteur pipette.
6. Wash three times with complete Ex-Cell medium supplemented with 2% FCS by centrifugation at $700 \times g$ for 10 min at room temperature.
7. Resuspend the cells in complete Ex-Cell medium supplemented with 10% FCS at concentration of 3×10^6/ml.
8. Add 10 ml of cell suspension to 75-cm^2 tissue culture flasks and incubate in CO_2 incubator at 37°C and 5% CO_2 for 5 days.

9. Remove the nonadherent cells and add 10 ml of fresh, complete Ex-Cell medium supplemented with 2% FCS.
10. Repeat Step 8 once or twice a week until the adherent cells reach approximately 85% confluence.
11. Passage using 0.5% w/v trypsin-EDTA solution or cultivate in complete Ex-Cell medium supplemented without FCS.

Comments

1. Be careful not to let cells reach 100% confluence, as they may detach spontaneously. These detached cells often do not reattach and must be discarded.
2. If you use fetal bone marrow cells, they grow faster and the final yield is approximately 60% higher.
3. Cryopreservation of bone marrow stromal cells is not recommended.

D. *Growth of human B cell precursors*

Materials and reagents

- Ex-Cell 300 medium (JRH Bioscience) supplemented with 50 μg/ml streptomycin and 50 U/ml penicillin
- X-VIVO serum-free medium (BioWhittaker)
- FCS
- Bone marrow stromal cells
- Human IL-7
- 96-well flat-bottom tissue culture plates
- Humidified CO_2 incubator

Protocol

(*Note:* All materials and reagents must be sterile and proper aseptic technique must be used when handling the cells.)

1. Prepare bone marrow stromal cells, as described above. Plate these cells in 96-well flat-bottom tissue culture plates in Ex-Cell medium with 10% FCS and antibiotics at 4×10^3 cells/well in 0.2 ml and incubate in CO_2 incubator at 37°C and 5% CO_2 for 5 days.
2. Remove the medium and replace with 0.2 ml of X-VIVO serum-free medium. Incubate in CO_2 incubator at 37°C and 5% CO_2 for at least 24 h.
3. Isolate and purify B cell precursors, as described previously. Resuspend them at concentration 5×10^5/ml in X-VIVO medium.
4. Prepare IL-7 in X-VIVO medium at 40 ng/ml.
5. Remove 0.1 ml of medium from wells with stromal cells and add 50 μl of B cell precursors suspension to each well.

6. Add 50 µl of IL-7 into each well.
7. Incubate in CO_2 incubator at 37°C and 5% CO_2.
8. Feed the cultures every 3 days by removing 100 µl of medium from each well and replacing with 100 µl of IL-7 (10 ng/ml) in X-VIVO medium.

Comments

1. Fetal calf serum is left out to inhibit further proliferation of stromal cells.
2. Stromal cells in serum-free medium can be used for 1 to 7 days after transfer.

E. Isolation of B lineage precursor cells

Magnetic purification of cell sorting based on cell cytoflourometer results in acceptable purity of individual B lineage precursor cells.

Materials and reagents

- Bone marrow cells isolated from fetal bone marrow
- Antibodies
- 100-mm petri dish
- 25-ml tissue culture flasks
- 15-ml conical centrifuge tubes
- Magnetic microspheres (Dynal; Miltenyi Biotech Inc.; Advanced Magnetics)
- RPMI 1640 medium supplemented with 10% FCS
- Humidified CO_2 incubator

Protocol

(*Note:* All materials and reagents must be sterile and proper aseptic technique must be used when handling the cells.)

1. Incubate cells in 25-ml tissue culture flasks for 2 h at 37°C in a humidified CO_2 incubator.
2. Wash two times with warm RPMI 1640 medium and collect nonadherent cells.
3. Incubate the nonadherent cells with appropriate antibodies (see Figure 1), as described in Chapter 1.
4. Positively or negatively isolate the desired population of B cell progenitors (Chapter 1).

Comments

1. Flasks from Step 1 can be used for isolation of stromal cells.

2. The cell sorting ability of a cell cytofluorometer equipped with this capacity can be used. The only substantial difference will be the use of fluorescently labeled antibodies.
3. The isolation of adherent cells also can be achieved by adherence to the Sephadex G-10 column (see Chapter 1).

F. *Mouse stromal cells*[31]

Materials and reagents

- IMDM serum-free medium (Gibco)
- RPMI 1640 medium supplemented with antibiotics and 2 *M* glutamine
- FCS
- Fetal liver cells
- Mouse or human IL-7
- 175-cm^2 tissue culture flasks
- 25-cm^2 tissue culture flasks
- 0.5% w/v Trypsin-EDTA solution
- Humidified CO_2 incubator

Protocol

(*Note:* All materials and reagents must be sterile and proper aseptic technique must be used when handling the cells.)

1. Isolate mouse bone marrow cells as described (Chapter 1). Cultivate the cells in RPMI 1640 medium supplemented with antibiotics and 10% FCS 175-cm^2 tissue culture flasks for 3 days in a CO_2 incubator at 37°C in 5% CO_2.
2. Remove nonadherent cells by discarding the supernatant. Add fresh medium and cultivate long enough for the cell to reach confluence.
3. Detach the cells by incubation in 0.5% w/v trypsin-EDTA solution.
4. Wash the cells once by centrifugation at $300 \times g$ for 10 min at 4°C.
5. Incubate 1×10^6 cells/175-cm^2 tissue culture flasks in 100 ml of complete RPMI 1640 medium in CO_2 incubator at 37°C in 5% CO_2.
6. Passage the cells approximately ten times; after ten passages the cells lose their growth capacity.
7. Isolate fetal liver cells as described (Chapter 1).
8. Incubate 8×10^5 fetal liver cells on the layer of murine stromal cells in 7 ml of RPMI 1640 medium with 200 U/ml of IL-7 in CO_2 incubator at 37°C in 10% CO_2.
9. Differentiation of pre-B cell clones is initiated by washing cells three times in IMDM serum-free medium supplemented with 2% FCS by centrifugation at $250 \times g$ for 5 min at 4°C.
10. Cultivate the cells in the same medium on a semiconfluent layer of stromal cells at a density of 1×10^7 cells in 7 ml medium without IL-7 in CO_2 incubator at 37°C in 10% CO_2.

11. After 2 days of incubation, cells are harvested and tested.

Comments

1. The most routinely used murine stromal cells are cell lines ST-2[32] and PA-6.[33] However, bone marrow–derived stromal cells can easily be used.
2. Adherent bone marrow cells usually reach confluence after 3 to 4 weeks of cultivation.
3. When stromal cell lines are used, these cells must be irradiated (30 gy) in γ-irradiator prior to using as feeder cells.
4. The recovery of viable cells is approximately 5% for normal pre-B cells and 85% for those expressing bcl-2 transgene.[34]

References

1. Owen, J. J. T., Raff, M. C. and Cooper, M. D., Studies on the generation of B lymphocytes in the mouse embryo, *Eur. J. Immunol.*, 5, 483, 1975.
2. Gathings, W. E., Lawton, A. R., and Cooper, M. D., Immunofluorescent studies of the development of pre-B cells, B lymphocytes and immunoglobulin isotope diversity in humans, *Eur. J. Immunol.*, 7, 804, 1977.
3. Kincade, P. W., Lee, G., Pietrangeli, C. E., Hayashi, S. I., and Gimble, J. M., Cells and molecules that regulate B lymphopoiesis in bone marrow, *Annu. Rev. Immunol.*, 7, 111, 1989.
4. Kincade, P. W. and Gimble, J. M., B lymphocytes, in *Fundamental Immunology*, Paul, W. E., Ed., Raven Press, New York, 1993, 43.
5. Kincade, P. W., Practical aspects of murine B lymphocyte cloning, in *Immunological Techniques Applied to Aging Research*, Adler, W. H. and Nordin, A. A., CRC Press, Boca Raton, 1991, 85.
6. Kincade, P. W. and Phillips, R. A., B lymphocyte development, *Fed. Proc.*, 44, 2874, 1985.
7. Witte, P. L., Kincade, P. W., and Větvička, V., Interculture variation and evolution of B lineage lymphocytes in long-term bone marrow culture, *Eur. J. Immunol.*, 16, 779, 1986.
8. Kincade, P. W., Defective colony formation by B lymphocytes from CBA/N and C3H/HeJ mice, *J. Exp. Med.*, 145, 249, 1977.
9. Větvička, V., Lee, G., and Kincade, P. W., Intrinsic B lymphocyte and macrophage defects in C3H/HeJ mice, *J. Immunol.*, 136, 2370, 1986.
10. Větvička, V., Different colony formation in low- and high-responding strains of mice, *Fol. Biol.*, 36, 130, 1990.
11. Hirayama, F. and Ogawa, M., Clonal culture assay for murine lymphohematopoietic progenitors, in *Immunology Method Manual CD-ROM*, Lefkovits, I., Ed., Academic Press, San Diego, 1997, 11.
12. Lemischka, I. R., Raulet, D. H., and Mulligan, R. C., Developmental potential and dynamic behaviors of hematopoietic cells, *Cell*, 45, 917, 1986.
13. Hirayama, F., Ship, J. P., Awgulewitsch, A., Warr, G. W., Clark, C. S., and Ogawa, M., Clonal proliferation of murine lymphohematopoietic progenitors in culture, *Proc. Natl. Acad. Sci. U.S.A.*, 89, 5907, 1992.
14. Baum, C. M., Weissman, I. L., Tsukamoto, A. S., Buckle, A. M., and Peault, B., Isolation of a candidate human hematopoietic stem-cell population, *Proc. Natl. Acad. Sci. U.S.A.*, 89, 2804, 1992.

15. Hirayama, F., Katayama, N., Neben, S., Donaldson, D., Nickbarg, E. B., Clark, S. C., and Ogawa, M., Synergistic interaction between interleukin-12 and steel factor in support of proliferation of murine lymphohematopoietic progenitors in culture, *Blood,* 83, 92, 1994.
16. Větvička, V., Kincade, P. W., and Witte, P. L., Effect of 5-fluorouracil on B lymphocyte lineage cells, *J. Immunol.,* 137, 2405, 1986.
17. Fernandez-Botran, R. and Větvička, V., *Methods in Cellular Immunology,* CRC Press, Boca Raton, 1995.
18. Kee, B. L. and Paige, C. J., Development of B lineage cells from uncommitted progenitors *in vitro,* in *Immunology Method Manual CD-ROM,* Lefkovits, I., Ed., Academic Press, San Diego, 1997, 18.
19. Cumano, A., Iscove, N. N., Paige, C. J., and Brady, G., Bipotential progenitors of B-cells and macrophages from fetal liver at day 12 of gestation, *Nature,* 356, 612, 1992.
20. Godin, I., Dieterlen-Lievre, F., and Cumano, A., Emergence of multipotent hematopoietic cells in the yolk sac and para-aortic splanchnopleura of 8.5 dpc mouse embryos, *Proc. Natl. Acad. Sci. U.S.A.,* 92, 773, 1995.
21. Kee, B. L., Cumano, A., Iscove, N. N., and Paige, C. J., Stromal cell independent growth of bipotent B-cell-macrophage precursors from murine fetal liver, *Int. Immunol.,* 6, 401, 1995.
22. Collins, L. S. and Dorshkind, K., A stromal cell line from myeloid long-term bone marrow cultures can support myelopoiesis and B lymphopoiesis, *J. Immunol.,* 138, 1082, 1987.
23. Dorshkind, K. and Landreth, K. S., Use of long-term bone marrow cultures and cloned stromal cell lines to grow B lineage cells, in *Immunology Method Manual CD-ROM,* Lefkovits, I., Ed., Academic Press, San Diego, 1997, 23.
24. Dexter, T. M., Allen, T. D., and Lajtha, L. G., Conditions controlling the proliferation of haematopoietic stem cells *in vitro, J. Cell. Physiol.,* 91, 334, 1977.
25. Whitlock, C. A. and Witte, O. N., Long-term cultures of B lymphocytes and their precursors from murine bone marrow, *Proc. Natl. Acad. Sci. U.S.A.,* 79, 3608, 1982.
26. Dorshkind, K., Regulation of hemopoiesis by bone marrow stromal cells and their products, *Annu. Rev. Immunol.,* 8, 111, 1990.
27. Dorshkind, K., *In vitro* differentiation of B lymphocytes from immature hemopoietic precursors present in long-term bone marrow cultures, *J. Immunol.,* 136, 422, 1986.
28. Pribyl, J. A., Shan, N., Dittel, B. N. and LeBien, T., Reagents for purification and growth of human B cell precursors in bone marrow stromal cell-dependent cultures, in *Immunology Method Manual CD-ROM,* Lefkovits, I., Ed., Academic Press, San Diego, 1997, 32.
29. Dittel, B. N., McCarthy, J. B., Wayner, E. A. and LeBien, T. W., Regulation of human B-cell precursor adhesion to bone marrow stromal cells by cytokines that exert opposing effects on the expression of vascular cell adhesion molecule-1 (VCAM-1), *Blood,* 81, 2272, 1993.
30. Wolf, M. L., Buckley, J. A., Goldfarb, A., Law, C. L., and LeBien, T., Development of a bone marrow culture for maintenance and growth of normal human B cell precursors, *J. Immunol.,* 147, 3324, 1991.
31. Rolink, A. and Melchers, F., *In vitro* growth and differentiation of early mouse B cell precursors, in *Immunology Method Manual CD-ROM,* Lefkovits, I., Ed., Academic Press, San Diego, 1997, 20.

32. Ogawa, M., Nishikawa, S., Ituka, K., Yamamura, F., Naito, M., Takahashi, K., and Nishikawa, S. I., B cell ontogeny in murine embryo studies by a culture system with the monolayer of a stromal cell clone, ST-2: B cell progenitor develops first in the embryonal body rather than in the yolk sac, *EMBO J.*, 7, 1337, 1988.
33. Kodama, H., Amagai, Y., Koyama, H., and Kasai, S., A new preadipose cell line derived from newborn mouse calvaria can promote the proliferation of pluripotent hemopoietic stem cells *in vitro, J. Cell. Physiol.*, 112, 89, 1982.
34. Rolink, A., Kudo, A., Karasuyama, H., Kikuchi, Y., and Melchers, F., Long-term proliferating early pre-B cell lines and clones with the potential to develop to surface Ig-positive, mitogen reactive B cells *in vitro* and *in vivo, EMBO J*, 10, 327, 1991.

chapter seven

Cell cultivation in vitro

Contents

Cultivation of cells *in vitro* belongs to the standard pool of techniques available not only in cellular immunology, but also in physiology, anatomy, histology, virology, molecular genetics, and biochemistry. In addition, the development of tissue culture as a modern technique helped to engender significant knowledge in the production of vaccines and in an understanding of neoplasia. In addition to its major advantages, such as the ability to control the environment, homogeneity of samples, and economy of cultivation, one must keep in mind the several disadvantages inherent in this technique. To cultivate the cells *in vitro*, the cells were dissociated from a three-dimensional network and from various types of intra-cellular cooperation. Thus, the validity of the cultured cells as a model of physiological function *in vivo* has often been criticized.

To better understand the problems connected with the propagation of cells *in vitro*, let us first offer some definitions. *Cell culture* means cells growing *in vitro* regardless of origin. These cells might originate when the original tissue is dispersed (either mechanically or enzymatically) into cell suspension, which is subsequently propagated. After an interval, the cells must be reseeded (due to their proliferation) into new flasks or wells. The most commonly used term is *passage*. Several subsequent passages resulting in cells with a similar growth capacity and a high degree of uniformity that form the *cell line*. It might be characterized by histological and/or biochemical means, but the major characteristic is its finite lifespan. The next step is the formation of an *established* or *continuous* cell line (with an infinite lifespan), usually by means of *transformation*.[1]

Cell cultivation is a technique demanding absolutely sterile working conditions. Media are rich in nutrients and thus provide excellent substrate for the growth of yeast and bacteria. In addition, cell cultures are quite sensitive to minor contaminations with trace chemicals or elements; therefore, careful attention must be focused on the cleaning and autoclaving of glassware. For detailed information about cell cultivation, including a cell culture laboratory setup, consult References 1 and 2.

I. Tissue culture laboratory

Essential requirements	*Desirable requirements*
Biohazard laminar flow	Separate biohazard laminar flow for animal dissection
Inverted microscope	Access to autoclave
Humidified CO_2 incubator	Centrifuge
Water bath	
Storage capacity in liquid nitrogen	

II. Aseptic techniques

Contamination by various microorganisms remains the biggest problem of cell cultivation *in vitro*. Risk of contamination can be minimized by following basic rules of aseptic treatment, regularly checking the cells under the microscope, prechecking all reagents for sterility *before* use, and not sharing any media with other people and/or different cell lines.

Basic rules of aseptic techniques:

1. Any material that comes into direct contact with the culture must be sterile.
2. Clean all work surfaces with 70% alcohol before and after manipulation with cells.
3. Mop up any spillage immediately.
4. Flame the necks of bottles and screw caps after opening and before closing bottles.
5. Do not leave bottles open for too long.

6. Do not pour medium from one sterile bottle into another. Always use a sterile pipette.
7. Change the filter in your laminar flow hood regularly, according to the manufacturer's instructions.
8. Ensure that your biosafety cabinet is certified annually.
9. If handling a culture known or suspected of carrying a human pathogen, wipe any spillage with a 5% formaldehyde.

III. Safety concerns

In addition to common concerns about safety, such as the handling of sharp items, the adequate handling of chemicals, or fire safety, the cultivation of cells *in vitro* includes another aspect. Biohazards are particularly important concerns, as we are routinely dealing with various pathogens. In addition, murine and human cell lines are contaminated with unknown viruses. Many cell lines have been transformed by viruses, but no epidemiological data are available for the significant assessment of potential risks. Because of possible hazards, the use of either a Class II Biosafety cabinet (a vertical laminar flow cabinet with front protection in the form of an air curtain) or a Class III Biosafety cabinet (a sealed pathogen cabinet with filtered air entering and leaving via a pathogen trap filter) is recommended. Potentially biohazardous material must be sterilized before disposal. The optimal procedure is to place this material into autoclavable sacks and to autoclave them promptly.

IV. Substrate

Glass flasks and dishes are economical, easily sterilized, and washable (and thus reusable), but their use is becoming less and less attractive. Many solutions commonly used in the laboratory are cytotoxic even in trace amounts. Glassware must be left soaking in water after use just to prevent the adherence of residual chemical components to the glass. For these reasons, disposable plastic material has become the material of choice for *in vitro* cell cultivation. For some cell cultures, it is beneficial to pretreat the surface of tissue culture plates. Cell growth is routinely improved by fibronectin or collagen.[3] There is no difference between the addition of extracellular factors or factors previously released from living cells.[4,5] One must bear in mind, however, that any type of surface pretreatment can induce specific alterations in some behavior (such as attachment) of the cells.[6]

V. Basic principles

The quality of water is extremely important. Only distilled and ionized water should be used in media and buffers preparation. Water coming from house purification systems, common in many laboratories, is often not purified enough, and further purification is highly recommended.

The mixture of penicillin (100 U/ml) and streptomycin sulfate (100 μg/ml) is the most commonly used combination of antibiotics. Despite the adequate effects of this mixture, some scientists prefer to use 30 μg/ml of gentamycin.

The addition of human or animal serum is necessary to sustain the optimal growth of cells *in vitro.* The most commonly used serum is fetal calf serum (FCS), a complex mixture of nutrients and other essential components. The optimal concentration and/or ratio of these components is still not completely clear. Therefore, for every culture application, FCS must be pretested for support of the cell line of interest. Most commercial sources are willing to offer free samples for testing and will put aside a much larger amount of a particular lot for later use. The optimal batch of FCS should be purchased in large quantities. Some types of experiments, such as cell activation and proliferation, require a much lower growth-potentiating activity of FCS. Most batches of FCS will, thus, be used in the laboratory. The use of heat-inactivated serum (see Chapter 1) is considered a standard procedure, even if there is no real proof that it is beneficial to cell cultures.

HEPES buffer is an important supplement in media used in an atmosphere without defined CO_2 content. A HEPES buffer used at concentration between 10 to 25 m*M* maintains the pH of the media regardless of atmospheric CO_2.

VI. Primary cultures

Generally, one of two different approaches can be used. One possibility is simply to let the cells migrate out of a small tissue fragment. The second option is to disaggregate the tissue (either enzymatically, mechanically, or both) before cultivation. There is no clearcut rule about which technique is preferred; it is recommended to try all possibilities. For mechanical disaggregation of tissue, a regular glass homogenizer is usually adequate and gentle enough. For enzymatical dissociation of tissue, the whole range of enzymes, often crude preparations only, have been used with varying degrees of success. Generally, trypsin and similar enzymes are very active, but they may cause rather extensive cell damage. On the other hand, collagenase is more expensive and less active, but also more gentle.

A. Basic rules of tissue disintegration

1. Carefully clean the tissue or organs and remove all necrotic tissue and/or fat.
2. Cut the tissue into very small pieces.
3. If the enzymes are used, remove them before cultivation.
4. Start with a higher cell concentration than usual; most of the cells will probably die.
5. Use the richest medium available with 20% FCS.

B. Disaggregation in trypsin

The mechanical and, even more, the enzymatic dissociation of cells not only yield a much higher number of cells, but also avoid problems with cell adherence. Note, however, that only cells resistant to mechanical or enzymatic disaggregation will be used for further propagation. This technique is particularly useful for embryotic tissue. With increasing age it becomes more difficult to obtain proliferating cells, mostly due to the decrease in numbers in the undifferentiated proliferating cell pool, increased amount of fibrous connective tissue, and/or extracellular matrix.

Materials and reagents

- Trypsin, diluted to 0.25% (Sigma; Difco; Gibco)
- PBS
- Tissue
- Magnetic bar
- 250-ml Erlenmeyer flask
- Magnetic stirrer
- 50-ml conical polystyrene centrifuge tube
- 100-ml petri dishes
- Forceps
- Scalpels
- 10-ml pipettes
- Hemocytometer
- RPMI 1640 medium supplemented with 10% FCS and antibiotics
- Humidified incubator with 5% CO_2
- 25-cm tissue culture flasks

Protocol

(*Note:* All materials and reagents must be sterile and proper aseptic technique must be used when handling the cells.)

1. Transfer tissue to the petri dish with RPMI 1640 medium and rinse.
2. Transfer the tissue to a second petri dish, dissect fat and necrotic material, and finely chop with a scalpel to about 1-mm pieces.
3. Transfer the small pieces into a 50-ml conical tube and add 40 ml of RPMI 1640 medium.
4. Gently shake, allow the tissue to settle down, and discard the medium.
5. Repeat Steps 3 and 4.
6. Transfer the chopped pieces to a 250-ml conical Erlenmeyer flask.
7. Add 100 ml of trypsin diluted in PBS and add a sterile magnetic bar.
8. Stir at 200 rpm for 30 min at 37°C.
9. Allow pieces to settle; collect supernatant.
10. Centrifuge three times at 500 × *g* for 5 min at 4°C.
11. Resuspend the pellet in 10-ml medium with 10% FCS.

12. Add fresh trypan to pieces and repeat Steps 7 to 11, until either complete disaggregation occurs or no further disaggregation is visible.
13. Pool all cell suspensions and count the cells.
14. Dilute the cells to 2×10^6 in RPMI 1640 medium with 10% FCS.
15. Transfer the cells into 25-cm tissue culture flasks.
16. Incubate the flasks in a humidified CO_2 incubator at 37°C. Check the flasks repeatedly for cell growth.
17. When the cell growth has spread over at least 50% of the surface, the cells may be passaged.

Comments

1. Wet the inside of the pipette before transferring the tissue fragments; otherwise, the pieces will stick to the pipette wall.
2. In order to be as gentle on cells as possible, it is recommended to collect the cells every 30 min of enzymatic treatment at 37°C.

C. Establishment of primary culture by cultivating tissue fragments

This technique is a modification of a technique originally developed by Harrison.[7] The finely chopped tissue is seeded into plates in a medium with a very high concentration of fetal calf serum.

Materials and reagents

- Tissue
- 100-ml petri dishes
- Forceps
- Scalpels
- 10-ml pipettes
- 50-ml conical polystyrene centrifuge tubes
- 25-cm tissue culture flasks
- RPMI 1640 medium supplemented with 50% FCS and antibiotics
- RPMI 1640 medium supplemented with antibiotics
- Humidified incubator with 5% CO_2

Protocol

(*Note:* All materials and reagents must be sterile and proper aseptic technique must be used when handling the cells.)

1. Transfer tissue to the petri dish with RPMI 1640 medium and rinse.
2. Transfer the tissue to a second petri dish, dissect fat and necrotic material, and finely chop with a scalpel into about 1-mm pieces.
3. Transfer the small pieces into a 50-ml conical tube and add 40 ml of RPMI 1640 medium.

4. Gently shake, allow the tissue to settle down, and discard the medium.
5. Repeat Steps 3 and 4.
6. Transfer the pieces into a tissue culture flask, about 20 pieces/flask.
7. Add 1 ml of RPMI 1640 medium supplemented with 50% FCS.
8. Incubate in a humidified incubator at 5% CO_2 for 24 h.
9. Check the flasks for adhesion of the pieces. If the pieces have adhered, you may gradually increase the volume of the medium, to approximately 5 ml over the next 4 days.
10. Change the medium weekly until a substantial outgrowth of cells is observed.
11. When the outgrowth has spread over at least 50% of the surface, the cells may be passaged.

Comments

1. The tissue pieces may be picked off from the flasks and transferred to a fresh flask with a fresh medium.
2. Wet the inside of the pipette before transferring the tissue fragments; otherwise, the pieces will stick to the pipette wall.
3. This method is particularly useful for small pieces of tissue which might be difficult to disaggregate (either enzymatically or mechanically).
4. This method is particularly suitable for adhering cell lines.

VII. Freezing of cell lines

Although commercial equipment for automated freezing of cell cultures exists, cell lines can be frozen without any specialized cell freezer. It is necessary to use a freezing rate of about 1°C/min, so the cells are not damaged by the ice crystal formation.

Materials and reagents

- Growing cell line
- Freezing vials (Dynatech; Nunc)
- Beaker
- Dry ice
- Styrofoam box
- RPMI 1640 medium supplemented with 10% FCS and antibiotics
- RPMI 1640 medium supplemented with 10% FCS, antibiotics, and 10% DMSO
- Pipettes
- 50-ml conical polystyrene tubes
- Centrifuge
- Thermometer

Protocol

(*Note:* All materials and reagents must be sterile and proper aseptic technique must be used when handling the cells.)

1. Wash the cells by centrifugation at 300 × *g* for 10 min at 4°C in RPMI 1640 medium.
2. Dilute cells at 1×10^7/ml.
3. Slowly (drop by drop) add the same volume of RPMI 1640 medium with 10% DMSO to the cell suspension.
4. Gently mix and aliquot the cell suspension into 1-ml cryogenic vials.
5. Immediately put the vials into a beaker and put the beaker into dry ice in a styrofoam box.
6. When the temperature inside the beaker has passed –50°C, you can transfer the vials into the liquid nitrogen long-term storage.

Comments

An even easier technique producing excellent results is routinely used in our laboratory. In this method, the cryogenic vials containing cells are put into the beaker and immediately transferred into the –80°C freezer. After overnight incubation, the vials are transferred into the –140°C freezer. A cell viability of over 75% is routinely found even after 7 years of storage.

References

1. Freshney, R. I., *Culture of Animal Cells. A Manual of Basic Technique,* Alan R. Liss, New York, 1987.
2. Horrigan, G., Condron, A., and Stanley, J. F., Housekeeping for cell culture laboratories, in *Laboratory Reagents in Immunology,* Vol. 1, Zola, H., Ed., CRC Press, Boca Raton, 1990, 3.
3. Hauschka, S. D. and Konigsberg, I. R., The influence of collagen on the development of muscle clones, *Proc. Natl. Acad. Sci. U.S.A.*, 55, 119, 1966.
4. Gilchrest, B. A., Nemore, R. E., and Maciag, T., Growth of human keratinocytes on fibronectin-coated plates, *Cell. Biol. Int. Rep.*, 4, 1009, 1980.
5. Stampfer, M., Halcones, R. G., and Hackett, A. J., Growth of normal human mammary cells in culture, *In Vitro* 16, 415, 1980.
6. Kleinman, H. K., McGoodwin, E. B., Rennard, S. I., and Martin, G. R., Preparation of collagen substrates for cell attachment: effect of collagen concentration and phosphate buffer, *Anal. Biochem.* 94, 308, 1981.
7. Harrison, R. G., Observation on the living developing nerve fiber, *Proc. Soc. Exp. Biol. Med.*, 4, 140, 1907.

chapter eight

Analysis of cytokine or cytokine receptor mRNA

Contents

I. Introduction

Cytokines are a group of soluble, regulatory glycoproteins secreted by cells of the immune system and other somatic cells. Cytokines normally act on cells of the immune system and other cells, inducing changes in their function, activation, and/or gene expression. It is now well accepted that cytokines function as key mediators of many physiologic and immunologic processes, including inflammation, immunity, and hemopoiesis.[1–4] Chemokines are a subgroup of cytokines whose main function is to act as "chemotactic" factors promoting the migration and recruitment of different types of leukocytes, such as neutrophils, monocytes, lymphocytes, and eosinophils.[3–6] The effects of cytokines on target cells are exerted by binding to specific membrane-bound cytokine receptors.[1–6]

During the last 10 years, numerous cellular and molecular studies have helped make cytokine biology one of the fastest growing areas of immunology. Currently, most murine and human cytokines and their receptors have been cloned; many cytokine-based therapies for human disease are in the experimental stages, and we possess a fairly good knowledge about the function of most major cytokines both in health and disease. Because most cytokines are produced only as a result of cellular activation by a relatively restricted spectrum of cell types, the assay of cytokines in biological samples has been widely used to monitor both inflammatory and immunological responses, as a correlate of cellular and immunological activation, and as means to identify different subsets of activated cells, particularly lymphocytes (i.e., Th1 and Th2 subsets of CD4 T cells).[7,8]

Although cytokines can be directly assayed by means of biological assays (based on their biological function using indicator cells or cell lines) or ELISAs,[9] some inherent problems are associated with these types of measurements. The first is the type of samples available. The assay of cytokine protein (or activity) is especially suited for liquid samples, such as tissue culture supernatants or biological fluids, but cannot be easily performed on solid tissue samples. The second is that the amount of cytokine in a biological sample represents basically a "spillover," or an excess over that cytokine amount that has been produced and utilized and/or degraded. Indeed, assays of cytokines, such as IL-2 or IL-4 in supernatants of cultures containing anti-IL-2R or anti-IL-4R antibodies to block their utilization, are considerably higher than in cultures without the anti-receptors antibodies, suggesting that the actual amounts of cytokine produced are much higher than normal measurements would indicate. Thus, cytokine measurements in supernatants or biological fluids may not always provide results that are representative of the "complete picture" concerning cytokine production.[10]

Despite the relatively more laborious nature of the techniques involved, analysis and quantitation of cytokine-specific mRNA levels have been employed to overcome some of the problems associated with the assay of cytokine protein or activity. The advantage of these types of assays is that they allow the analysis of the expression of a particular cytokine or group of cytokines by a defined tissue, organ, or group of cells at any given time. Moreover, the sensitivity of some of the techniques used to determine cytokine mRNAs is very high, allowing the study of cytokine expression on a small number of cells.[10–12] In contrast, the sensitivity of ELISAs may sometimes be insufficient. Several techniques have been employed to analyze and quantitate cytokine mRNAs, each with its own advantages and disadvantages.[13] Among the most commonly used are Northern blotting,[14] ribonuclease protection assay (RPA),[15] and reverse transcriptase-polymerase chain reaction (RT-PCR).[10–13] These techniques will be briefly examined, with special attention to RT-PCR, following a discussion on the isolation of RNA.

II. Total RNA isolation

The isolation of RNA is the first and most critical step in the analysis of the expression of cytokine/cytokine receptor mRNAs.[10–13] Failure to obtain good quality RNA in this step often leads to errors and irreproducible results later on. An essential consideration is the stability of the isolated RNA molecules, which are very sensitive to degradation by ribonucleases, present both in the original samples and as a practically ubiquitous contaminant of laboratory solutions and equipment. Therefore, the key to obtaining good RNA preparations is to minimize ribonuclease activity during cell/tissue lysis and initial extraction steps, coupled with an avoidance of the introduction of trace ribonucleases from contaminated glassware or solutions. As part of these precautions, all glassware or plasticware must be autoclaved and handled with gloves; water used to prepare solutions should be treated with diethyl-pyrocarbonate (DEPC), an RNAse-inhibitor and, whenever possible, all equipment and plasticware to be used for RNA isolation should not be used for other purposes.[16]

RNA extraction procedures use the potent chaotropic agent, guanidinium isothiocyanate, and reducing agents such as 2-mercaptoethanol to disintegrate cellular structures, dissociate nucleoproteins and, at the same time, inactivate endogenous RNAses.[17] Most of the commonly used protocols of RNA extraction follow the basic technique described by Chomczynski and Sacchi,[18] which uses a single-step, acid guanidinium thiocyanate-phenol-chloroform mixture that is easily adaptable to microfuge tubes. The basic technique and reagents are described in this chapter. A number of commercially prepared reagents and kits for RNA extraction are available. When using one of these, readers are urged to follow the manufacturer's instructions.

Materials and reagents

- Acid guanidinium thiocyanate-phenol solution (e.g., RNA STAT-60, Tel-Test; RNAwiz, Ambion)
- Chloroform (ACS or Molecular Biology grade, e.g., Sigma)
- Isopropanol (ACS or Molecular Biology grade)
- Ethanol (ACS or Molecular Biology grade)
- Double distilled water, DEPC-treated (see Chapter 13)
- Microfuge tubes (1.5 ml, autoclaved)
- Pipette tips (barrier tips recommended)
- Microcentrifuge
- Refrigerator
- Freezer (–80°C)

Protocol

1. For cell suspensions, lyse cells directly in the guanidinium thiocyanate-phenol reagent. Pass the lysate several times through a pipette or needle in order to aid disruption. Use approximately 1 ml of reagent for up to 5×10^6 cells (in a 1.5-ml microfuge tube). For adherent cell monolayers, the reagent can be added directly onto the cells. Avoid washing the cells, as this increases the chance of RNA degradation.

2. For tissue samples, place approximately 50 to 100 mg of tissue in 1 ml of the guanidinium thiocyanate-phenol reagent, using a microfuge (1.5 ml-) tube. Homogenize with a glass-Teflon® or Polytron homogenizer.
3. Incubate the homogenate/lysate for 5 min at room temperature to allow for the dissociation of nucleoprotein complexes.
4. Add 0.2 ml of chloroform to each tube, cap tightly, and shake vigorously for 15 s. Let the tubes stand for 2 to 3 min at room temperature.
5. Centrifuge the homogenate at 12,000*g* for 15 min at 4°C. After centrifugation, the homogenate will separate into two phases: a lower organic phase (chloroform) and an upper aqueous phase, which contains the cellular RNA. Proteins and genomic DNA separate into the organic phase or the interphase.
6. Transfer the aqueous phase to a clean microfuge tube and add 0.5 ml of isopropanol. *Be careful to avoid the interphase!* (Contamination with protein/DNA will occur if this step is not carried out carefully.)
7. Precipitate the RNA by incubating samples for at least 10 min at 4°C. Isolation of RNA from homogenates prepared from low numbers of cells or small tissue samples might benefit from a longer incubation at this step (e.g., overnight).
8. Centrifuge at 12,000*g* for 10 min at 4°C. The RNA should be visible as a white pellet at the bottom of the tube, unless starting with a small number of cells.
9. Carefully aspirate the supernatant, resuspend the pellet in 1 ml of a 75% ethanol solution, vortex, and centrifuge at 7500*g* for 5 min at 4°C.
10. Carefully aspirate the supernatant, invert the microfuge tube, and allow the RNA pellet to dry (5 to 10 min), although not completely. Do not use the Speed-Vac for this purpose.
11. Resuspend the RNA pellet in DEPC-treated water containing 1 m*M* EDTA and vortex, or pipette repeatedly through a pipette tip in order to solubilize the RNA completely.
12. Quantitate RNA by measuring $OD_{260/280}$ in a UV-spectrophotometer. An OD reading of 1 at 260 nm corresponds approximately to 40 μg/ml of RNA. The $OD_{260/280}$ ratio gives an indication of the purity of the preparation. Ideally, ratios between 1.8 and 2.0 are desirable, as they are indicative of good RNA purity. Contamination by proteins or phenol will significantly decrease the $OD_{260/280}$ ratio. Samples might have to be diluted for OD reading.
13. If not used immediately, store samples at –80°C. If in doubt, the integrity of the RNA preparation can be ascertained by agarose-formaldehyde gel electrophoresis.[16]

Comments

1. Depending on the cell type, 1×10^6 cells should yield approximately 5 to 15 μg of RNA. Depending on the tissue type, 1 mg of tissue will yield 1 to 10 μg RNA.

2. Cell or tissue homogenates in the acid guanidinium-phenol reagent can be stored frozen at –80°C for at least 2 weeks before processing.
3. Messenger RNA (mRNA) species can be isolated after this step by affinity chromatography using oligo(dT)-cellulose columns. However, most procedures for the analysis of cytokine mRNA expression (e.g., RT-PCR, RPA) do not require further purification of the RNA.
4. DEPC-treated water is prepared by adding DEPC to double-distilled water to a final concentration of 0.1% w/v. Allow to stand overnight at room temperature and autoclave to remove residual traces of DEPC. Do not treat either Tris or ammonium-containing solutions with DEPC. *Important:* DEPC is a suspected carcinogen; therefore, take appropriate precautions.

III. Reverse transcription-polymerase chain reaction (RT-PCR)

After the original introduction of the polymerase chain reaction (PCR), which allowed the amplification of small amounts of target DNA,[19] this technique was adapted for the detection of RNA by including an initial reverse transcription step in which cDNA is synthesized for its later use as a template in the PCR amplification step.[20] The PCR reaction uses sets of specific primers that allow amplification of target sequences. By making use of the known sequence information on most cytokines and chemokines (both of human and animal origin), oligonucleotide primers can be easily designed and synthesized (sequence information can be obtained from cytokine handbooks or the Internet, including Web sites for the National Center for Biotechnology Information (NCBI) [http://www.ncbi.nlm.nih.gov] and the European Bioinformatics Institute (EBI) [http://www.ebi.ac.uk]). The cytokine-specific primers allow the amplification of the extremely low amounts of the generated cytokine cDNA up to levels that can be easily detected by a variety of techniques, such as ethidium bromide staining of agarose gels (after electrophoresis); by Southern blotting and probing with an internal oligo DNA probe; or by colorimetric techniques using probes labeled with fluorescent dyes, such as digoxigenin. Kits that adapt this latter technique with a 96-well format have become available recently (e.g., Quantikine®, R&D Systems). The sensitivity of the RT-PCR technique has made it the method of choice for the detection of specific RNA molecules in very small numbers of cells or RNA molecules that are normally expressed at very low levels.

Although a negative or positive signal has been adequate enough for some experimental purposes, many studies require quantitation or, at the very least, comparison of the relative levels of cytokine mRNAs in different experimental groups. Even though the concentration of the amplified product should be, in theory, directly proportional to the amount of starting cDNA, sample-to-sample variations in RNA stability and recovery, or technical errors, are amplified many times, thus making quantitation relatively tricky and cumbersome.[13] Many immunological studies making use of RT-PCR techniques to study cytokine expression have actually employed semi-quantitative reagents, based on the amplification and comparison of the cytokine mRNA sequence of choice and a "control" housekeeping gene, whose steady-state mRNA levels are known (or assumed) to be constant under the conditions of the experiment (e.g., actin, hypoxanthine phosphoribotransferase [HPRT], glyceraldehyde-3-phosphate dehydrogenase [G3PDH]).[21] These serve to control variations in RNA recovery and loading, as well as the efficiency of the RT and PCR reactions. Results are often expressed as the ratios of the gene-specific signal to that of the internal control.

True quantitation in RT-PCR measures the absolute amount (e.g., copy number) of a specific mRNA sequence in a sample. Quantitative RT-PCR techniques employ a competitive strategy, in which a known amount of a synthetic RNA (amplifiable using the same set of primers) is added to the samples. The RNA competitor yields PCR products of sizes different from the specific sequence being measured, thus allowing identification. Two basic approaches have been used for the preparation of the internal standards for competitive RT-PCR, including the cloning of a modified construct ("competitor") generated by creating a deletion in the endogenous target region (therefore of a smaller size); and the synthesis of a construct with identical primer binding sites but unrelated internal sequence ("mimics").[10,13] Accurate quantitation requires that the exogenous standard be amplified with the same efficiency as the endogenous target. Finally, the PCR product from the endogenous transcript is compared and quantitated based on a standard concentration curve created by the exogenous RNA.

Table 1 includes some examples of the sequence of primers used for the analysis of selected cytokines; however, sequence information for the construction of primers for the analysis of human or mammalian cytokines/chemokines is readily available via the Internet. In addition, primer sets and oligonucleotide probes for many human and mouse cytokines can be purchased from a number of vendors. For quantitative RT-PCR, commercial kits are available for the preparation of "competitor" RNA (e.g., RT-PCR Competitor Construction kit, Ambion). Kits employing a colorimetric detection and quantitation of cytokine PCR products are also available, as mentioned previously.

The modified semi-quantitative RT-PCR procedure of Wynn et al.[22] is described in this chapter.

A. *Reverse transcriptase step*

Materials and reagents

- Deoxynucleotide triphosphate mixture (dNTPs), containing 2.5 m*M* each of dATP, dGTP, dCTP, dTTP (Boehringer-Mannheim; #1051-440, 458, 466, and 482)

Table 1 Sequences of Probes for Cytokine RT-PCR

		Size (b.p.)
	Mouse	
IL-1α	S 5′-CTCTAGAGCACCATGCTACAGAC-3′	308
	A 5′-TGGAATCCAGGGGAAACACTG-3′	
	P 5′-TGTAAGAATACCCAGACAGCTTTAAGGATGGGAGGG-3′	
IL-1β	S 5′-TGAAGGGTGCTTCCAAACCTTTGACC-3′	322
	A 5′-TGTCCATTGAGGTGGAGAGCTTTCAGC-3′	
	P 5′-AGAAGAGCCCATCCTCTGTGACTCAGTGGA-3′	
IL-2	S 5′-TGATGGACCTACAGGAGCTCCTGAG-3′	167
	A 5′-GAGTCAAATCCAGAACATGCCGCAG-3′	
	P 5′-CACCTTCAAATTTTACTTGCCCAAGCAGGCC-3′	
IL-3	S 5′-GTGGCCGGGATACCCACCGTTTAAC-3′	
	A 5′-TGGCAGCGCAGAGTCATTCGCAGAT-3′	
IL-4	S 5′-ACGAGGTCACAGGAGAAGGGACGCCATGCA-3′	188
	A 5′-TCATTCATGGAGCAGCTTATCGATGAATCC-3′	
	P 5′-CTCACAGCAACGAAGAACACCACAGAGAGT-3′	
IL-6	S 5′-GTGACAACCACGGCCTTCCCTACT-3′	352
	A 5′-GGTAGCTATGGTACTCCA-3′	
	P 5′-CACAGAGGATACCACTCCCAACAGACC-3′	
IL-10	S 5′-TCCTTAATGCAGGACTTTAAGGGTTACTTG-3′	240
	A 5′-GACACCTTGGTCTTGGAGCTTATTAAAATC-3′	
	P 5′-CGGCTGAGGCGCTGTCATCGATTTCTCCCC-3′	
IL-12p40	S 5′-CTGCCACAAAGGAGGCGAGACCTC-3′	
	A 5′-CATATTTATTCTGCTGCCGTGCTTC-3′	
TNFα	S 5′-GCGACGTGGAACTGGCAGAAG-3′	277
	A 5′-GGTACAACCCATCGGCTGGCA-3′	
	P 5′-CAGTTCTATGGCCCAGACCCTC-3′	
IFNγ	S 5′-TGGAGGAACTGGCAAAAGGATGGT-3′	336
	A 5′-TTGGGACAATCTCTTCCCCAC-3′	
	P 5′-GTGGACCACTCGGATGAGCTCATT-3′	
TGFβ	S 5′-AGACGGAATACAGGGCTTTCGATTCA-3′	
	A 5′-CTTGGGCTTGCGACCCACGTAGTA-3′	
HPRT	S 5′-GTAATGATCAGTCAACGGGGGAC-3′	214
	A 5′-CCAGCAAGCTTGCAACCTTAACCA-3′	
	P 5′-GCTTTCCCTGGTTAAGCAGTACAGCCCC-3′	
β-actin	S 5′-GGACTCCTATGTGGGTGACGAGG-3′	
	A 5′-GGGAGAGCATAGCCCTCGTAGAT-3′	
	Rat	
IL-1β	S 5′-CCAGGATGAGGACCCAAGCA-3′	519
	A 5′-TCCCGACCATTGCTGTTTCC-3′	
IL-2	S 5′-GCGCACCCACTTCAAGCCCT-3′	351
	A 5′-CCACCACAGTTGCTGGCTCA-3′	
IL-4	S 5′-TGACGAGCAATGAGACGATG-3′	275
	A 5′-TTTCAGTGTTCTGAGCGTGGA-3′	
IL-5	S 5′-TGACGAGCAATGAGACGATG-3′	229
	A 5′-TCATCACGCCAAGGAACTCT-3′	

Table 1 (continued) Sequences of Probes for Cytokine RT-PCR

		Size (b.p.)
IL-6	S 5′-CTTCCAGCCAGTTGCCTTCT-3′	496
	A 5′-GAGAGCATTGGAAGTTGGGG-3′	
IL-10	S 5′-TGCCTTCAGTCAAGTGAAGACT-3′	346
	A 5′-AAACTCATTCATGGCCTTGTA-3′	
TNFα	S 5′-CGAGTGACAAGCCCGTAGCC-3′	468
	A 5′-GGATGAACACGCCAGTCGCC-3′	
IFNγ	S 5′-CCCTCTCTGGCTGTTACTGC-3′	419
	A 5′-CTCCTTTTCCGCTTCCTTAG-3′	
TGFβ_1	S 5′-GCCTCCGCATCCCACCTTTG-3′	396
	A 5′-GCGGGTGACTTCTTTGGCGT-3′	
HPRT	S 5′-TCCCAGCGTCGTGATTAGTG-3′	608
	A 5′-GGCTTTTCCACTTTCGCTGA-3′	
β-actin	S 5′-CTATCGGCAATGAGCGGTTC-3′	762
	A 5′-CTTAGGAGTTGGGGGTGGCT-3′	
	Human	
IL-1β	S 5′-TACGAATCTCCGACCACCACTACAG-3′	295
	A 5′-TGGAGGTGGAGAGCTTTCAGTTCATATG-3′	
IL-2	S 5′-ACTCACCAGGATGCTCACAT-3′	266
	A 5′-AGGTAATCCATCTGTTCAGA-3	
IL-4	S 5′-CCTCTGTTCTTCCTGCTAGCATGTGCC-3′	373
	A 5′-CCAACGTACTCTGGTTGGCTTCCTTCA-3′	
IL-6	S 5′-AGCTCAGCTATGAACTCCTTCTC-3′	338
	A 5′-GTCTCCTCATTGAATCCAGATTGG-3′	
IL-10	S 5′-ATGCCCCAAGCTGAGAACCAAGACCCA-3′	352
	A 5′-TCTCAAGGGGCTGGGTCAGCTATCCCA-3′	
IL-12p40	S 5′-CCAAGAACTTGCAGCTGAAG-3′	355
	A 5′-TGGGTCTATTCCGTTGTGTC-3′	
IL-13	S 5′-TATGCATCCGCTCCTCAATCCTC-3′	449
	A 5′-CGAAGTTTCAGTTGAACCGTCC-3′	
TNFα	S 5′-CGGGACGTGGAGCTGGCCGAGGAG-3′	354
	A 5′-CACCAGCTGGTTATCTCTCAGCTC-3′	
IFNγ	S 5′-AGTTATATCTTGGCTTTTCA-3′	356
	A 5′-ACCGAATAATTAGTCAGCTT-3′	
β-actin	S 5′-ATTGCCGACAGGATGCAGAA-3′	150
	A 5′-GCTGATCCACATCTGCTGGAA-3′	
	P 5′-CAAGATCATTGCTCCTCCTGAGCGCA-3′	

- Reverse transcriptase (RT) buffer 5× (250 m*M* Tris-HCl, pH 8.3, 375 m*M* KCl, 15 m*M* $MgCl_2$) (supplied with Reverse transcriptase, Gibco BRL, #18053)
- Dithiothreitol (DTT), 0.1 *M* solution
- RNasin (Promega, # N2512)
- Random hexamer oligonucleotides, diluted to 20 mA_{260}/μl (Boehringer-Mannheim, #1034-731)
- Superscript Reverse Transcriptase (Life Technologies BRL, #18053-017)

- DEPC-treated double-distilled water
- RNase-free microfuge tubes (0.5 ml)
- Pipettes
- RNase-free pipette tips (barrier tips recommended)
- Thermocycler
- Microcentrifuge

Protocol

1. Determine the number of samples to be run and prepare enough of the following mixture (amounts given are per sample): 2.5 µl deoxynucleotide phosphates mixture, 5.0 µl Reverse transcriptase buffer, 2.0 µl DTT solution (0.1 M), 0.5 µl RNasin solution, 2.0 µl random hexamer oligonucleotides (20 units/µl), and 1.0 µl Superscript reverse transcriptase. Aliquot 13 µl of the mixture in RNase-free microfuge tubes.
2. Using RNase-free microfuge tubes, mix enough sample (approximately 1 µg RNA) and DEPC-treated water to a final volume of 12 µl. Heat the tubes to 70°C for 5 min and immediately put on ice for 5 min. Quickly spin the tubes (2 to 3 s) in a microcentrifuge and let sit at room temperature for another 5 min.
3. Add the RNA samples to the reaction tubes, vortex, and centrifuge quickly (2 to 3 s) to bring the mixture to the bottom of the tube. Incubate at 37°C for 60 min to allow first-stand cDNA synthesis. Terminate enzyme activity by incubating at 90°C for 5 min and place the tubes on ice for 10 min. Use the thermocycler to run this protocol.
4. Dilute cDNA with 175 µl DEPC-treated water. At this point, samples can be stored indefinitely at –20°C.

B. Polymerase chain reaction

Materials and reagents

- Deoxynucleotide triphosphates (dNTPs) mix (2.5 m*M*)
- Taq polymerase, 5 U/µl (Promega, #M-1862)
- PCR Buffer (supplied together with Taq polymerase)
- Primers (sense and antisense mixture 1:1) 0.2 µ*M* final
- DEPC-treated water containing 1.5 m*M* $MgCl_2$
- Mineral oil (Molecular Biology grade)
- RNase-free microfuge tubes (0.5 ml)
- Thermocycler
- Microcentrifuge

Protocol

1. Using RNase-free microfuge tubes, mix 10 µl of sample (from RT reaction) with 4.0 µl dNTP mixture, 5.0 µl PCR buffer, 2.0 µl primer

mix, 0.2 µl Taq polymerase, and 28.8 µl 1.5 m*M* $MgCl_2$ solution. Overlay mixtures with 50 µl of mineral oil.

2. Place tubes in thermocycler and carry out polymerase chain reaction under the following conditions:
   (a) 1 cycle at 95°C for 30 s (initial denaturation)
   (b) 25 to 35 cycles (optimal number of cycles need to be determined in pilot experiments) at 94°C for 1 min (denaturation), 54°C for 1 min (primer annealing), and 72°C for 2 min (extension)
   (c) 1 cycle at 72°C for 7 min (final extension)
3. At this step, the PCR products are ready for quantitation and can be stored indefinitely at –20°C.

C. Detection of PCR products

After amplification, PCR products can be separated by agarose or polyacrylamide gel electrophoresis and directly visualized by ethidium-bromide staining, or transferred to nylon membranes (Southern blotting) and probed with labeled oligonucleotide probes. Alternatively, PCR products can be visualized by dot-blot hybridization techniques, without electrophoresis and Southern blotting (assuming the identity of the products has been confirmed in pilot experiments).[21] The latter technique, however, cannot be used in competitive PCR, as the products derived from the endogenous and exogenous standard sequences need to be differentiated based on their size. The methodology described here is based on the identification of PCR products by Southern blotting and chemoluminiscent detection.[22]

1. Southern blot analysis

Materials and reagents

- Electrophoresis grade Agarose (e.g., Seakem LE Agarose, FMC Bio-Products)
- Tris-borate EDTA buffer (TBE; 0.089 *M* Tris, 0.089 *M* boric acid, 2 m*M* EDTA, pH 8.0)
- Sample loading buffer 6× (40% w/v sucrose, 0.25% bromophenol blue, 0.25% xylene cyanol)
- Ethidium bromide (EtBr)
- Molecular size markers (e.g., 1 kb ladder)
- 1.5 *M* NaCl, 0.5 *M* NaOH (denaturing solution)
- 1.5 *M* NaCl, 1*M* Tris-HCl, pH 7.5 (neutralizing solution)
- 10× SSC solution (1.5 *M* NaCl, 0.17 *M* sodium citrate, pH 7.0)
- Nylon membrane (e.g., Hybond N$^+$, Amersham Pharmacia Biotech)
- Horizontal gel apparatus
- Electrophoresis power supply
- UV transilluminator

- Vacuum blotting system (optional)
- UV crosslinker
- Vacuum oven
- Polaroid camera and film
- Parafilm
- Paper towels

Protocol

1. Remove 13.2 µl from the PCR amplified samples from below the mineral oil. Wipe the pipette tip to remove any oil. Place on a piece of Parafilm.
2. To the sample on the Parafilm, add 4.8 µl of 6× gel loading buffer. Mix.
3. Load samples into a 1% agarose gel containing EtBr (0.5 µg/ml). The size of the gel depends on the size of the electrophoresis apparatus and the number of samples to be analyzed. Load one track of the gel with molecular size markers (e.g., 1 kb ladder) in loading buffer.
4. Run the gel (using TBE as electrophoresis buffer) at 150 V (for a large gel) for approximately 20 min.
5. Visualize the DNA bands using a transilluminator. Wear eye and face protection. A record of the gel can be made at this point using Polaroid film or via digital-image capture technology.
6. Denature the gel by soaking in 200 ml of a 1.5 *M* NaCl, 0.5 *M* NaOH solution for 30 to 60 min, with constant stirring or shaking.
7. Remove denaturing solution and rinse gel briefly with distilled water. Add 200 ml of neutralizing solution (1.5 *M* NaCl, 1*M* Tris-HCl, pH 7.5) and soak the gel for at least 30 min.
8. Meanwhile, cut a nylon membrane (e.g., Hybond N^+, Amersham Pharmacia Biotech) exactly to the size of the gel. Wet the membrane in 10× SSC solution.
9. Transfer the DNA from the gel to the membrane. This can be done with the help of an electrophoretic or vacuum blotting apparatus (e.g., VacuGene XL, Amersham). Transfer time is approximately less than 1 h using the vacuum system. Alternatively, transfer can be accomplished without using any special equipment by sandwiching the gel and the membrane between paper towels and two glass plates, placing them in a tray containing 10× SSC (up to the level of the bottom plate), and then applying a weight (~500g) on top of the glass plate. This allows a flow of buffer from the reservoir through the gel and the membrane, so DNA fragments are deposited on the latter. A detailed description of this method is found in Maniatis et al.[16]
10. Remove membrane and mark as necessary.
11. Cross-link the DNA to the membrane using a UV crosslinker, and then bake at 80°C for 2 h in a vacuum oven. Membranes can now be hybridized or stored in a dessicator.

2. Hybridization and detection

The procedure described here for hybridization and detection is based on the 3′-labeling of oligonucleotide probes with fluorescein and detection by chemiluminiscence. This procedure uses the ECL 3′-oligolabelling and detection system from Amersham (#RPN2131). Alternatively, probes can be labeled with ^{32}P, alkaline phosphatase, or fluorescent compounds and detected accordingly.

Materials and reagents

- ECL 3′-oligolabelling and detection system (Amersham #RPN2131)
- Oligonucleotide probes
- 5× SSC solution containing 0.1% SDS (hybridization buffer)
- Hybridization oven
- Imaging film (e.g., Kodak BioMax MR-1; Eastman Kodak)
- Darkroom
- Pipettes
- Saran Wrap

Protocol

1. Label oligonucleotide probe with fluorescein, using the reagents and the instructions provided by the manufacturer in the ECL 3′-oligolabelling kit. Labeled probes can be stored at –20°C for at least 6 months.
2. Prehybridize blots in hybridization oven at 42°C for a minimum of 30 min in a 5× SSC - 0.1% SDS solution containing the blocking agent (Follow manufacturer's instructions.)
3. Add 7.5 ng of labeled probe per ml of hybridization solution (10 ml is usually enough for an average-sized blot of 20 × 5 cm). Place the blot with the hybridization solution in a polystyrene bag, seal, and incubate overnight at 42°C.
4. Carefully remove the blot from the bag and wash twice with 5× SSC - 0.1% SDS solution for 5 min at room temperature. Wash two more times with a 1× SSC - 0.1% SDS solution for 15 min at 45°C with continuous shaking.
5. Develop blots according to the instructions provided by the manufacturer and expose the film. Exposure times are rarely longer than 10 min. Be sure to obtain several exposures to ensure that the resulting signal is within the linear range of the film.
6. Scan and quantitate appropriate bands in the films, using a densitometer or computer scanner.
7. Results are expressed as the ratio of the signal of the cytokine mRNA being measured to that of the internal housekeeping control.

3. RNase protection assay (RPA)

The ribonuclease protection assay is a very sensitive and specific technique for both detection and quantitation of mRNA species.[15] It involves the hybridization of a radioactively labeled antisense RNA probe to the target mRNA in solution, followed by exposure to RNases that degrade the free probe and any unhybridized or single-stranded RNA. The remaining "protected" species are then separated on a polyacrylamide gel and visualized and quantitated by autoradiography and densitometry or phosporimaging techniques. The intensity of the corresponding bands is directly related to the original amount of target RNA in the sample.

Although quantitation in the RPA is much more straightforward than that for the RT-PCR, it did not enjoy the wide popularity of the latter for the analysis of cytokine mRNAs, mostly because the available methodologies made it time- and labor-consuming to analyze the expression of several cytokine mRNAs in a sample (it meant running one reaction for each cytokine-specific mRNA probe). In the last few years, however, commercially available kits have allowed the simultaneous analysis of multiple cytokines (up to ten) in the same sample (RiboQuant, Pharmingen/Becton Dickinson). These multi-probe RPA kits are available in different panels of probes (e.g., Th1- or Th2-cytokines, pro-inflammatory cytokines, chemokines, etc.) for the analysis of human, mouse, and rat cytokines, and cytokine receptors. Conveniently, the analysis of cytokine or cytokine receptor mRNAs by the multi-probe RPA kits does not require purification of poly-A$^+$ RNA, as the total RNA preparations described earlier in this chapter are sufficient.

This technique will not be covered in more detail here as a complete set of instructions, including the labeling of probes, hybridization, and RNAse digestion, is provided by the manufacturer.

References

1. Arai, K., Lee, F., Miyajima, A., Miyatake, S., Arai, N., and Yokota, T., Cytokines: coordinators of immune and inflammatory responses, *Annu. Rev. Biochem.*, 59, 783, 1990.
2. Oppenheim, J. J. and Sklatvala, J., Cytokines and their receptors, in *Clinical Applications of Cytokines: Role in Pathogenesis, Diagnosis and Therapy*, Oppenheim, J. J., Rossio, J., and Gearing, A., Eds., Oxford University Press, New York, 1993, 3.
3. Abbas, A. K., Lichtman, A. H., and Pober, J. S., *Cellular and Molecular Immunology*, 2nd ed., W. B. Saunders, Philadelphia, 1994, 240.
4. Janeway, C. A. and Travers, P. *Immunobiology. The Immune System in Health and Disease*, 2nd ed., Current Biology/Garland, London, 1998.
5. Baggiolini, M., Dewald, B., and Moser, B., Human chemokines: an update. *Annu. Rev. Immunol.*, 15, 675, 1997.
6. Kim, C. H. and Broxmeyer, H. E., Chemokines: signal lamps for trafficking of T and B cells for development and effector function, *J. Leuk. Biol.*, 65, 6, 1999.

7. Rossio, J. L., and Gearing, A. J. H., Measurement of cytokines, in *Clinical Applications of Cytokines: Role in Pathogenesis, Diagnosis and Therapy,* Oppenheim, J. J., Rossio, J., and Gearing, A., Eds., Oxford University Press, New York, 1993, 16.
8. Mosmann, T. R. and Coffman, R. L., Th1 and Th2 cells: different patterns of lymphokine secretion lead to different functional properties, *Annu. Rev. Immunol.,* 7, 145, 1989.
9. Fernandez-Botran, R. and Větvička, V., *Reagents in Cellular Immunology,* CRC Press, Boca Raton, 1995, 77.
10. Vandebriel, R. J., Van Loveren, H., and Meredith, C., Altered cytokine (receptor) mRNA expression as a tool in immunotoxicology, *Toxicology,* 130, 43, 1998.
11. O'Garra, A. and Vieira, P., Polymerase chain reaction for detection of cytokine gene expression, *Curr. Opinion Immunol.,* 4, 211, 1992.
12. Walker, K. B., Detection and analysis of cytokine mRNA in tissues and cell lines, *J. Immunol. Reagents,* 212, 113, 1998.
13. Delassus, S., Quantification of cytokine transcripts using polymerase chain reaction, *Eur. Cytokine Netw.,* 8, 239, 1997.
14. Alwine, J. C., Kemp, D. J., and Stark, G. R., Method for detection of specific RNAs in agarose gels by transfer to diazobenzyloxy-methyl paper and hybridisation with DNA probes, *Proc. Natl. Acad. Sci. U.S.A.,* 74, 5350, 1977.
15. Gilman, M., Ribonuclease protection assay, in *Current Protocols in Molecular Biology,* Vol. 1, Ausubel, F. M., Brent, R., Kingston, R. E., Moore, D. D., Seidman, J. G., Smith, J. A., and Stuhl, K., Eds., John Wiley & Sons, New York, 1993, 4.7.1.
16. Maniatis, T., Fritsch, E. F. and Sambrook, J., *Molecular Cloning — A Laboratory Manual,* 2nd ed., Cold Spring Harbor Laboratory Press, New York, 1989.
17. Chirgwin, J. M., Przybyla, A. E., MacDonald, R. J., and Rutter, W. J., Isolation of biologically active ribonucleic acid from sources enriched in ribonuclease, *Biochemistry,* 18, 5294, 1979.
18. Chomczynski, P. and Sacchi, N., Single step method of RNA isolation by acid guanidinium thiocyanate-phenol-chloroform extraction, *Anal. Biochem.,* 162, 156, 1987.
19. Saiki, R. K., Gelfand, D. H., Stoffel, S., Scharf., S. J., Higuchi, R., Horn, G. T., Mullis, K. B., and Erlich, H. A., Primer-directed enzymatic amplification of DNA with a thermostable DNA polymerase, *Science,* 239, 487, 1988.
20. Krug, M. S. and Berger, S. L., First-strand cDNA synthesis primed with oligo(dT), *Reagents Enzymol.,* 152, 316, 1987.
21. Murphy, E., Hieny, S., Sher, A., and O'Garra, A., Detection of *in vivo* expression of interleukin-10 using a semi-quantitative polymerase chain reaction method in *Schistosome mansoni* infected mice, *J. Immunol. Reagents,* 162, 211, 1993.
22. Wynn, T. A., Eltoum, I., Cheever, A. W., Lewis, F. A., Gause, W. C., and Sher, A., Analysis of cytokine mRNA expression during primary granuloma formation induced by eggs of *Schistosoma mansoni, J. Immunol.,* 151, 1430, 1993.

chapter nine

Intracellular cytokine staining

Contents

Because of the central involvement of cytokines in the regulation of practically all immunological and inflammatory processes, assays for the detection and quantitation of cytokines have been extensively used, both for basic and clinical research purposes. Initially, cytokines were assayed in culture supernatants or biological fluids based on their biologic activity using *in vitro* target cell lines (bioassays).[1,2] However, although these assays were very sensitive, they often suffered from relative poor specificity, as most target cell lines could respond to more than one cytokine. The subsequent availability of anti-cytokine mAbs allowed more specific quantitation of cytokines, based on the capture-and-detection immunoassays such as ELISAs, in wide use today.[2] Moreover, molecular biology techniques such as ribonuclease protection assays (RPA) and reverse transcriptase-polymerase chain reaction (RT-PCR) have allowed detection and quantification of cytokine-specific mRNAs.[3–5]

One limitation to the above techniques is their "bulk" nature. That is, they often measure cytokine production of mRNA expression in complex, heterogenous cell populations, and thus are unable to provide information on the identity and frequency of cells producing a particular cytokine in the whole cell population.[6,7] Moreover, in the absence of pure cell preparations, these techniques are not well suited for the study of cytokine production by defined cell subsets, such as in the case of the Th1 and Th2 subsets of CD4 T cells.[7] Inasmuch as available evidence supports the notion that changes in the frequency of different cell subsets occur in several diseases, the ability to detect cytokine production by individual, phenotypically defined cells has become particularly important.

Initially, techniques such as limiting dilution analysis and ELISPOT were used to determine the frequency of cells producing a particular cytokine in a heterogenous cell population.[8] These methodologies, however, are relatively time consuming and labor intensive. Recently, techniques developed by Jung et al.[9] and Picker et al.,[10] based in the intracellular staining of cytokines, have tremendously simplified the study of cytokine production at the single cell level. In addition to their remarkable specificity and sensitivity, these techniques are not affected by factors that normally interfere with the "bulk" cytokine assays, particularly the presence of membrane and soluble cytokine receptors.

As its name implies, this technique is based on the detection of intracellular cytokine molecules with fluorescent-labeled anti-cytokine antibodies, following short-term activation (4 to 6 h) of the cells with a known stimulus in the presence of a blocker of protein transport, such as Brefeldin A or Monensin.[6,7,9] These inhibitors cause the accumulation of cytokine molecules inside the cytoplasm of the cells. The cells are also stained with antibodies against membrane markers (e.g., CD3, CD4, CD8) in order to study defined cell populations. Permeabilization of the cell membrane with non-ionic detergents or alcohols (e.g., saponin) is required in order to allow the detection of the intracellular cytokine molecules.[6,7,9]

The basic general technique is described below. There are many variables in this assay, particularly regarding the origin and preparation of the cells (e.g., tissue culture, peripheral blood, lymphoid organs, etc.) and the stimulatory conditions. These are described elsewhere.[2] Also, depending on the cell population to be analyzed and the cytokines under investigation, the antibodies used for the detection of cell markers and cytokines will vary. Great care should be invested in the planning of the cell isolation and activation protocol and in the selection of the most appropriate antibodies for the purposes of the experiment. Many antibodies are available from commercial sources already conjugated to a variety of fluorochromes. In addition, antibodies and reagents specifically designed for intracellular cytokine staining protocols can also be obtained commercially (FastImmune™ Cytokine System, Becton Dickinson).

I. Cell activation

Materials and reagents

- Cell population (harvested from tissue culture, whole blood, PBMC, lymphoid organ suspensions)
- Disposable 12 × 75-mm polystyrene tubes or equivalent
- Vortex mixer
- Refrigerated centrifuge
- Incubator (37°C, 5 to 7% CO_2)
- Pipettors

- Brefeldin A (Sigma, B-7651). Prepare a 5-mg/ml stock solution in DMSO. Aliquot and freeze.
- Monensin (Sigma, M-5273)

Protocol

1. Prepare cell population to be studied according to appropriate procedure.
2. Wash and resuspend the cells in culture medium to a density of 1 to 2 $\times 10^6$ per ml. Culture the cells in the presence of the appropriate activation stimulus. For example, a combination of phorbol myristate acetate (PMA, 25 ng/ml) and ionomycin (1 μg/ml) are routinely used for the activation of lymphocytes. Other agents may include mitogens, superantigens, or activating antibodies. Appropriate kinetic experiments need to be performed to determine the optimal activation time (4 to 6 hours for PMA + ionomycin). Brefeldin A (10 μg/ml) or monensin (2 μ*M*) are included in the culture for the final 4 to 5 h of activation.
3. Control unstimulated cultures in the presence of Brefeldin A should also be included.
4. At the end of the culture period, harvest cells and wash them once with ice-cold HBSS.

II. Staining

Materials and reagents

- Activated cell population
- Disposable 12 × 75-mm polystyrene tubes or equivalent
- Vortex mixer
- Refrigerated centrifuge
- Pipettors
- Fluorochrome-labeled antibodies
- Staining buffer (HBSS containing 0.5% BSA)
- Paraformaldehyde (Sigma, P-6148). Prepare a 2% solution in PBS.
- Saponin (Sigma, S-4521). Prepare a 0.1% w/v solution in staining buffer.

Protocol

1. Aliquot cells (1 to 5 $\times 10^{-5}$/test) and transfer to tubes.
2. Stain the cells for surface markers by incubating with the appropriate labeled antibodies (at 2 μg/ml in staining buffer) for 30 min at 4°C in a final volume of 0.1 ml. For example, anti-CD3-PerCP antibodies are useful to gate on $CD3^+$ T cells.
3. Wash the cells twice with staining buffer. Spin down at 500 × *g*. Fix the cells by adding 0.5 ml of a 2% paraformaldehyde solution in PBS to the cell pellet. Incubate for 20 min at room temperature.

4. Wash the cells and permeabilize with 0.5 ml of a 0.1% saponin solution in staining buffer. Incubate for 20 min at room temperature.
5. Wash twice and resuspend the cell pellets in 100 μl of staining buffer.
6. Add the fluorescent-labeled anti-cytokine mAbs (2 to 4 μg/ml final concentration). Incubate for 30 min on ice.
7. Wash twice and resuspend the cell pellets in staining buffer containing 1% paraformaldehyde.
8. Analyze in FACS flow cytometer. Acquire data with CellQuest™ or LYSIS II™ software, using a fluorescence or forward scatter (FSC) threshold.
9. Analyze data using CellQuest, LYSIS II, PAINT-A-GATE™, or Attractors™ software.

Comment

Negative controls for each antibody used (isotype-matched irrelevant antibodies) should be included.

References

1. Fernandez-Botran, R. and Větvička, V., *Reagents in Cellular Immunology,* CRC Press, Boca Raton, 1995, 77–146.
2. Rossio, J. L. and Gearing, A. J. H., Measurement of cytokines, in *Clinical Applications of Cytokines: Role in Pathogenesis, Diagnosis and Therapy,* Oppenheim, J. J., Rossio, J., and Gearing, A., Eds., Oxford University Press, New York, 1993, 16.
3. Delassus, S., Quantification of cytokine transcripts using polymerase chain reaction, *Eur. Cytokine Netw.,* 8, 239, 1997.
4. Murphy, E., Hieny, S., Sher, A., and O'Garra, A., Detection of *in vivo* expression of interleukin-10 using a semi-quantitative polymerase chain reaction method in *Schistosome mansoni* infected mice, *J. Immunol. Reagents,* 162, 211, 1993.
5. Walker, K. B., Detection and analysis of cytokine mRNA in tissues and cell lines, *J. Immunol. Reagents,* 212, 113, 1998.
6. Prussin, C. and Walker, D. D., Detection of intracytoplasmic cytokine using flow cytometry and directly conjugated anti-cytokine antibodies, *J. Immunol. Reagents,* 188, 117, 1995.
7. Maino, V. C. and Picker, L. J., Identification of functional subsets by flow cytometry: intracellular detection of cytokine expression, *Cytometry,* 34, 207, 1998.
8. Carter, L. L. and Swain, S. L., Single cell analyses of cytokine production, *Curr. Opin. Immunol.,* 9, 177, 1997.
9. Jung, T., Schauer, U., Heusser, C., Neumann, C., and Rieger, C., Detection of intracellular cytokines by flow cytometry, *J. Immunol. Reagents,* 159, 197, 1993.
10. Picker, L. J., Singh, M. K., Zdraveski, Z., Treer, J. R., Waldrop, S. L., Bergstresser, P. R., and Maino, V. C., Direct demonstration of cytokine synthesis heterogeneity among human memory/effector T cells by flow cytometry, *Blood,* 86, 1408, 1995.

chapter ten

Cell cytotoxicity

Contents

Cellular cytotoxicity is among the most important functions of the cellular branch of defense reactions. Since the time the first assays were developed, there has been a continuing controversy regarding the actual mechanisms of cell-mediated cytotoxicity. This controversy persists today, despite decades of work and much real progress. One must bear in mind that there are several different types of cellular cytotototoxicity, which differ fundamentally from one another.

Basic techniques of cell-mediated cytotoxicity have been described in a previous volume.[1] This book focuses on new techniques dealing with all three aspects of cellular toxicity, i.e., natural killer cell cytotoxicity, antibody-dependent cell-mediated cytotoxicity (ADCC), and T-cell cytotoxicity (CTL). The physiological function of CTLs is not quite clear, but it is a common belief that they control the virus infection by killing virus-infected cells.

The discovery of secretory granules in NK cells and CTLs, with subsequent demonstration of perforin and cytolysis, suggest the means of the cytotoxic effects.[2,3] The granulocyte exocytosis model of cytotoxicity is appealing because it can explain most of the experimental data in terms of our knowledge of cell biology. The current concept of CTL cytotoxicity suggests that an initial target-killer cell contact is followed up by a Ca^{2+}-dependent lethal hit and subsequent lysis step without any further involvement of the killer cell. Electron microscopy studies of CTLs and NK cells have shown that the membrane of killer cells becomes polarized and preexisting granules are secreted into the narrow gap between the killer cells and their prey.[3] Subsequent studies have measured the release of soluble granule components after delivery of a trigger stimulus.[4] The trigger stimulus must be a polyvalent stimulus such as immobilized ligand binding and it is Ca^{2+} dependent. Perforin binds quite effectively to membranes with widely spaced phospholipids. Besides perforin, a variety of esterases have been identified as essential for the lytic activity.[5] The current hypothesis suggests that granzyme A, which is an esterase colocalized with perforin, penetrates into target cells with the help of perforin, and induces DNA fragmentation,[6] probably by nucleolin cleaving. The target cell death is accompanied by several characteristic events, including apoptosis, DNA fragmentation, and fatty acid production.[7]

I. Preparation of mouse cytotoxic T cells[8]

Materials and reagents

- Mice
- Responder cell population (see Chapter 1)
- RPMI 1640 medium supplemented with antibiotics, 25 m*M* HEPES, and 10% FCS
- RPMI 1640 medium
- Concanavalin A (12 µg/ml in complete RPMI 1640 medium)
- Mouse rIL-2
- Phytohemagglutinin (PHA, Difco)
- 96-well V-bottom microtiter plates
- Humidified CO_2 incubator

Protocol

(*Note:* All materials and reagents must be sterile and proper aseptic technique must be used when handling the cells.)

1. Prepare the responder cell population from the tissue of interest.
2. Irradiate the donor mice at 1200 rad and allow mice to rest 60 min.
3. Sacrifice the mice and prepare spleen cell population (feeder cells), as described in Chapter 1.
4. Resuspend the responder cells to appropriate dilutions (10, 20, 50, and 100 cells/well) and feeder cells (10^7 cells/ml) at complete RPMI 1640 medium.

5. Put 50 µl of feeder cells and 50 µl of responder cells into each well of a 96-well V-bottom microtiter plate.
6. Add 50 µl of diluted Concanavalin A and 25 U/ml rIL-2 into each well of a 96-well V-bottom microtiter plate.
7. Incubate plates for 8 days in a humidified CO_2 incubator at 37°C.

Comments

1. The test of their ability to lyse target cells in a lectin-facilitated assay is recommended.
2. The optimal dose for irradiation of mice should be determined in a preliminary experiment.
3. As an alternative, splenocytes might be irradiated directly (3500 rad). There are no real differences between these two options, but in some facilities it might not be feasible to irradiate living animals.

II. Whole blood assay[9]

Clinical studies often require frequent monitoring of NK activity. However, in some patients, including small children, the quantity of blood available is severely restricted. The following procedure offers an assay of a sensitive whole blood technique using only 0.6 ml of peripheral blood without any subsequent isolation procedures. The results are comparable to the standard NK assay.

Materials and reagents

- Human peripheral blood in heparin
- K562 cells as target cells
- $Na_2{}^{51}CrO_4$ for cell labeling (see Chapter 1)
- Round-bottom 96-well plates with cover (Costar)
- Tubes suitable for your γ-counter
- Pipettes
- γ-Counter
- 37°C humidified CO_2 incubator
- Centrifuge

Protocol

1. Add 200 µl of undiluted blood into wells of the round-bottom 96-well plate.
2. Pipette 50 µl of ^{51}Cr-labeled K562 (diluted to 1×10^4/ml) target cells into each well.
3. Do not forget to leave empty wells as blanks. Wells with effector cells only and target cells only should also be prepared.

4. Incubate the plate at 37°C in a humidified 5% CO_2 incubator for 18 h.
5. Centrifuge the plates for 5 min at 550 × *g* at room temperature. Carefully transfer the supernatant into small tubes optimal for your γ-counter.

Comments

1. The samples should be used in triplicate.
2. Use only heparin as an anticoagulant agent. Acid-citric dextrose or EDTA gives unsatisfactory results.[9]
3. Due to the lower numbers of NK cells in peripheral blood, an incubation shorter than 18 h is not recommended.

III. DNA fragmentation[10]

Several independent observations showed that target cells attacked by cytotoxic lymphocytes undergo morphologic changes similar to those occurring during programmed cell death, termed apoptosis. The mechanism of such cell death was based on induced fragmentation of target cell DNA that progresses to the nucleosomal level within minutes of target-killer cell contact. The degree of DNA solubilization was shown to be dependent upon the nature of the target cells. To describe these differences, some hypotheses suggested that the killer cells induce the same damage to the nuclear envelope in all targets, but that this damage leads to nuclear disintegration in only some of them. However, later experiments found no evidence that the nuclear envelope is damaged by CTLs in target cell types that do not undergo nuclear disintegration.[11]

The true verification of DNA fragmentation requires agarose gel electrophoresis to reveal the classical ladder pattern. For quantification of DNA fragmentation, the evaluation of the soluble fraction is necessary. The basic principle of this technique is that DNA that has suffered extensive double-strand breakage is rendered soluble. At the same time, intact DNA remains insoluble. Recently, the use of ^{3}H-thymidine over more commonly used $^{125}IUdR$ has been recommended. Among the suggested advantages are (a) a typical short pulse of $^{125}IUdR$ labels only a small fraction of DNA; (b) it labels only cells in S phase; (c) radioactive iodine is about 100 times more cytotoxic than ^{3}H, so the isotope might be the primary source of DNA damage.

Materials and reagents

- Allospecific cytotoxic lymphocytes
- Target cells
- 75-cm^2 polystyrene tissue culture flasks
- RPMI 1640 medium supplemented with 10% FCS, and antibiotics
- ^{3}H-Thymidine (New England Nuclear; 20 μCi/ml)

- 37°C humidified CO_2 incubator
- 12 × 75-mm polystyrene test tubes
- Concanavalin A (Sigma)
- Vortex
- DNase I (Sigma)
- Centrifuge
- Water bath
- Triton X-100
- Rabbit complement (Accurate Chemical and Scientific Corp.)
- Scintillation vials
- Scintiverse Cocktail (Fisher)
- Liquid scintillation counter

Protocol

1. Incubate 5×10^6 CTLs in 25 ml of RPMI 1640 medium in a 75-cm^2 polystyrene tissue culture flask with 250 µl of ^{3}H-thymidine overnight in a 37°C humidified CO_2 incubator.
2. Wash the cells five times by centrifugation in warm RPMI 1640 medium at 150 × *g* at room temperature.
3. Add 100 µl of target cells (2.5×10^5/ml), 100 µl of CTLs (E/T depends on type of target), and 100 µl of Concanavalin A at a final concentration of 5 µg/ml to 12 × 75-mm tubes.
4. Vortex the tubes and centrifuge them at 500 × *g* for 3 min at room temperature.
5. Incubate the tubes in a 37°C water bath for 2 h.
6. Add the following in this order: 500 µl of 0.2% Triton X-100, 100 µl of rabbit complement, and 100 µl of DNase.
7. Vortex the tubes and incubate them in a 37°C water bath for 60 min.
8. Centrifuge the tubes at 1300 × *g* for 15 min at 4°C.
9. Transfer the supernatant into scintillation vials.
10. Add 3 ml of Scintiverse Cocktail (or another scintillation liquid) and count in a liquid scintillation counter.

Comments

1. To accurately determine the percentage of lysed cells, DNA fragmented prelytically within the unlysed cells must not be soluble, i.e., Triton X-100 must not be used.
2. Entire DNA of every lysed cell must be soluble; therefore, DNase is added prior to harvest.[10]
3. Rabbit complement is not used when some targets (such as Raji cell line) are used. In such a case, 100 µl of FCS is used instead.

IV. Apoptosis

Apoptosis is physiologically regulated or even preprogrammed cell death in which the chromatin becomes condensed and the DNA is degraded. It is important to remember that this type of cell death occurs under normal physiological conditions. The term "programmed" or "preprogramed" death refers to the fact that it is one of the normal cell functions. Apoptotic processes are accompanied by loss of membrane integrity, which makes it rather simple to quantitate. Apoptotic cells do not swell, but contract.[12] However, changes in nuclear morphology, such as compaction of nuclear chromatin into dense masses, are more important.[12–14] Agarose gel electrophoresis of DNA from apoptotic cells shows a distinctive ladder-like pattern, which has become a typical sign of apoptosis. The next step in the apoptotic events is the actual breakup of cells resulting in cell death. The remains of the dead cells are phagocytosed by surrounding tissue. The mechanisms controlling the processes of apoptosis in cells of the immune system are complex. For a review see References 15 and 16.

V. Quantitation of apoptosis using fluorescent dyes[16]

The use of DNA-binding fluorescent dye is a simple technique allowing us to determine the percentage of cells undergoing apoptosis. The combination of ethidium bromide and acridine orange is the combination of choice, as other DNA-binding dyes require less common excitation wavelengths (<350 nm). Acridine orange stains DNA green. RNA is also stained, but as acridine orange cannot intercalate into RNA (as it does with DNA), RNA is stained orange. Dead cells will be stained with ethidium bromide (living cells can exclude this dye) and will appear orange.

Materials and reagents

- Cell suspension
- RPMI 1640 medium supplemented with 10% FCS and antibiotics
- 100 μg/ml acridine orange (Sigma) in PBS
- 100 μg/ml ethidium bromide (Sigma) in PBS
- 12 × 75-mm glass tubes
- Microscope slides and coverslips
- Pipettes
- Fluorescence microscope

Protocol

1. Dilute cells to 1×10^6 cells/ml in complete RPMI 1640 medium.
2. Place 25 μl of cell suspension in bottom of the tube. Add 1 μl acridine orange and 1 μl ethidium bromide and mix gently.

3. Place 10 μl of this suspension on the microscopic slide. Cover with coverslip and examine under the fluorescence microscope with appropriate filters.
4. Count 200 to 300 cells.
5. Apoptotic index and percentage of dead cells can be determined.

Comments

1. Both acridine orange and ethidium bromide are mutagenic; proper care is necessary.
2. Four different subpopulations of cells will appear:
   (a) green with chromatin with organized structure, i.e., living cells with no apoptosis
   (b) green with apoptotic (condensed or fragmented) nuclei, i.e., apoptotic cells
   (c) orange with chromatin with organized structure, i.e., dead cells
   (d) orange with apoptotic (condensed or fragmented) nuclei, i.e., cells dead by apoptosis

VI. Quantitation of DNA fragmentation[16]

This technique is based on the fact that fragmented DNA does not sediment together with chromosome-length DNA when subjected to centrifugation.[14] This technique is significantly more complicated than the previous technique without any real advantages. It is recommended mainly for evaluation of cell populations that cannot be easily labeled in their DNA, e.g., resting lymphocytes.

Materials and reagents

- Cell suspension
- RPMI 1640 medium supplemented with 10% FCS and antibiotics
- 5% and 25% Trichloracetic acid
- Diphenylamine reagent (see Chapter 1); prepare fresh before use
- Microcentrifuge
- Tris buffer containing EDTA (see Chapter 1) and 0.2% Triton X-100
- 96-well flat-bottom microtiter plates
- 1.5-ml microcentrifuge tubes
- Vortex
- ELISA reader

Protocol

1. Dilute cells to 1×10^6 cell/ml in complete RPMI 1640 medium.
2. Add 0.5-ml cell suspension into microcentrifuge tube labeled A.

3. Pellet cells by centrifugation at 200 × *g* for 10 min at 4°C.
4. Transfer supernatant into a new centrifuge tube labeled B.
5. Add 0.5 ml Tris buffer containing EDTA and 0.2% Triton X-100 to cell pellet (tube A) and vortex.
6. Centrifuge tube A for 10 min at 4°C at maximum speed.
7. Transfer the supernatant into a new centrifuge tube labeled C.
8. Add 0.5 ml Tris buffer containing EDTA and 0.2% Triton X-100 to cell pellet in tube A.
9. Add 0.5 ml of 25% trichloracetic acid to each tube and vortex thoroughly.
10. Incubate overnight at 4°C.
11. Precipitate DNA into pellet by centrifugating 10 min at 4°C at maximum speed.
12. Aspirate and discard the supernatant.
13. Add 80 µl of 5% trichloracetic acid to each tube and hydrolyze DNA by heating 15 min at 90°C. Do not forget blank tube with only 80 µl of 5% trichloracetic acid.
14. Add 160 µl of diphenylamine reagent (prepared immediately before use) to each tube and vortex thoroughly.
15. Incubate overnight at room temperature.
16. Transfer 200 µl of solution into individual wells of a 96-well plate and read optical density at 600 nm using an ELISA reader.
17. Calculation of percent fragmented DNA:

$$\%\ \text{fragmented DNA} = \frac{B+C}{A+B+C} \times 100$$

Comments

1. A spectrophotometer can be used instead of an ELISA reader. In such a case, the colored solution will be transferred directly into the cuvets.
2. A range of excitation wavelengths between 520 nm and 620 nm can be used.

VII. Target cell detachment assay[17]

This technique is aimed at evaluation of the functional consequences resulting from T cell-tumor interaction. In addition to measuring detachment of target cells – T cells, it also simultaneously measures lytic activity.

Materials and reagents

- Adhering target cells (macrophages, fibroblasts)
- Effector cells

- $Na_2{}^{51}CrO_4$ for cell labeling (see Chapter 1)
- RPMI 1640 medium supplemented with 10% FCS and antibiotics
- PBS with 1 m*M* EDTA
- 96-well flat-bottom microtiter plates (Immulon 2, Dynatech)
- Pipettes
- Pasteur pipettes
- Centrifuge
- 37°C humidified CO_2 incubator
- 0.2% Triton X-100
- Tubes suitable for your γ-counter
- γ-Counter
- Vortex

Protocol

1. Seed adherent target cells in wells of a 96-well flat-bottom microtiter plate at 2×10^4 cells/well in 200 µl of RPMI 1640 medium.
2. Incubate for 24 h before the assay.
3. Wash each well three times by aspiration of medium and adding 150 µl of warm RPMI medium.
4. Radiolabel monolayers by adding 50 µl of warm medium containing 200 µCi/ml of ^{51}Cr. Incubate for 2 h at 37°C.
5. Wash the wells five times by aspiration of medium and addition of 150 µl of warm RPMI medium.
6. Add 2×10^5 effector cell/well in 200 µl of RPMI 1640 medium and centrifuge the plate at $450 \times g$ for 2 min.
7. Incubate the plate for 18 h at 37°C.
8. Prepare control wells containing target cells only with and without 100 µl 0.2% Triton X-100.
9. Place a Pasteur pipette against the side of the well with a tip near, but not touching, the bottom. Vigorously pipette the cell suspension six times, and collect the medium and nonadherent cellular fraction (dead and live cells) into tube containing 1 ml PBS with 1 m*M* EDTA.
10. Vortex the cell mixture.
11. Centrifuge at $500 \times g$ for 5 min at 4°C. Transfer one half of the supernatant to a fresh tube.
12. Quantitate the radioactivity in both tubes and calculate the target cell detachment and lysis.

Comments

1. If macrophages are used as targets, seed 4×10^4 cells/well.
2. Target cell detachment reflects intact as well as damaged cells and is determined as the sum of radioactivity in both tubes divided by the total radioactivity incorporated into the monolayer.

VIII. Cytolysin hemolytic activity[18]

Materials and reagents

- 0.2% (w/v) PEG 6000 (Sigma) in PBS
- Human red blood cells in heparin
- Cytolysin (or isolated cytolytic granules)
- Hanks' balanced salt solution containing 0.01 *M* HEPES and 2 mg/ml bovine serum albumin
- 0.1% Triton X-100
- 37°C humidified CO_2 incubator
- Pipettes
- Centrifuge
- 96-well round-bottom microtiter plates (Costar)
- 96-well flat-bottom microtiter plates (Immulon 2, Dynatech)
- ELISA reader

Protocol

1. Dilute human red blood cells to 0.2% (v/v) suspension in Hanks' solution containing HEPES and BSA.
2. In a round-bottom plate, add 100 µl of PBS with PEG to a row of wells for each sample. To the first well add cytolysin and buffer to a volume of 200 µl.
3. Mix and carry out twofold serial dilutions of samples (the final volume in each well will be 100 µl).
4. Add 100 µl of diluted erythrocytes into each well and incubate the plates at 37°C for 10 min.
5. Spin the plates at 500 × *g* for 5 min.
6. Remove 150 µl of supernatant and transfer it to a flat-bottom plate.
7. Read the absorbance at 420 nm using an ELISA reader.

Comments

1. Do not forget to prepare control wells with PBS-PEG only and with 0.1% Triton X-100.
2. Numerous substances (such as membrane fragments, lipoprotein) were shown to inhibit cytolysin activity.[19] Negative results may thus mean inhibitor activity masking the presence of cytolysin.
3. The addition of PEG helps prevent adsorption of highly diluted proteins to the plastic.

IX. Isolation of cytoplasmic granules[20]

Studies of isolated and purified cytoplasmic granules provided much better insights into mechanisms of cellular cytotoxicity than observation of *in vitro* killing activity.

Materials and reagents

- Isolated NK cells or cultivated large granular lymphocytes (LGLs)
- Hanks' balanced salt solution
- Disruption buffer: PBS containing 0.25 *M* sucrose, 0.01 *M* HEPES, 4 m*M* EGTA and 1×10^3 U/ml heparin (Sigma), pH 7.4
- Disruption buffer: PBS containing 0.25 *M* sucrose, 0.01 *M* HEPES, and 4 m*M* EGTA, pH 7,4
- $MgCl_2$
- DNAse I (Sigma)
- Nucleopore filters (Corning)
- Percoll (Pharmacia)
- Ultracentrifuge

Protocol

1. Resuspend NK cells at 1×10^8/ml in disruption buffer.
2. Lyse the cells at 0°C.
3. After addition of $MgCl_2$ to 5 m*M*, digest the homogenate with 800 U/ml DNAse I for 15 min at 22°C.
4. Remove nuclei by filtration through 5 and 3 μm Nucleopore filters.
5. Cool the homogenate to 0°C.
6. Layer 5 ml aliquots on 20 ml of 48% Percoll in disruption buffer without heparin.
7. Centrifuge at 29,000 × *g* in an ultracentrifuge. Collect the purified granules from the visible band near the bottom of the tubes. Remove remaining Percoll by centrifugation of the granule fraction at 85,000 × *g* for 16 h at 0°C.

X. Esterase assay[21]

Serine esterases are among the numerous substances found to be active in cell toxicity.[22] Although the exact functional role of these enzymes in the process is still rather controversial, the assay based on quantization of esterases released during CTL activation is simple and reliable. Cytotoxic lymphocytes can be activated by several different approaches: by monoclonal antibodies mimicking the effect of antigen-bearing target cells, by a combination of calcium ionophore and protein kinase C (or phorbol ester), or by traditional incubation with target cells.

XI. Activation by antibodies[21]

Materials and reagents

- Cytotoxic T lymphocytes (CTL)
- PBS

- Monoclonal antibodies against T cell receptor complex, 5 μg/ml in PBS
- RPMI 1640 medium supplemented with 10% FCS and antibiotics
- 1% (v/v) Triton X-100
- 20 m*M* *N*-α-benzyloxycarbonyl-L-lysine thiobenzyl ester (Calbiochem)
- 22 m*M* 5,5′-dithio-*bis*(2-nitrobenzoic acid) (Pierce Chemicals)
- 0.1 *M* phenylmethanesulfonyl fluoride (Sigma) in dimethylsulfoxide (Sigma)
- 96-well round-bottom microtiter plates (Dynatech)
- 37°C humidified CO_2 incubator
- Water bath
- 12 × 75-mm polypropylene culture tubes
- Centrifuge
- 50-ml conical centrifuge tubes
- Ice bath
- Spectrophotometer

Protocol

1. Add 50 μl of anti-T cell receptor complex antibody to each well of the 96-well round-bottom microtiter plate.
2. Incubate the plate for 30 min at room temperature.
3. Discard the liquid from each well and wash the wells three times with complete RPMI 1640 medium.
4. Add 50 μl of CTL (1×10^7/ml) in complete RPMI 1640 medium to each well. Add another 50 μl of complete RPMI 1640 medium to each well, except those used for the determination of total granule release. Add 40 μl of complete RPMI 1640 medium and 10 μl of 1% Triton X-100 into those wells.
5. Incubate the plates for 4 h in a 37°C humidified CO_2 incubator.
6. Centrifuge the plates at 200 × *g* for 5 min at 4°C.
7. Transfer 50 μl of supernatant to 12 × 75-mm polypropylene culture tubes.
8. Prepare substrate solution:
   500 μl 20 m*M* *N*-α-benzyloxycarbonyl-L-lysine thiobenzyl ester
   500 μl 22 m*M* 5,5′-dithio-*bis*(2-nitrobenzoic acid)
   500 μl 1% Triton X-100
   48.5 ml PBS
9. Add 950 μl of substrate solution to supernatant aliquots and incubate at 37°C in water bath.
10. Move plates to an ice bath.
11. Immediately add 1 μl of 0.1 *M* phenylmethanesulfonyl fluoride to each tube.
12. Add 1 ml of PBS to each tube.
13. Measure the absorbance of samples at 412 nm.
14. Calculate percentage values using the following formula:

$$\% \text{ secretion} = \frac{\text{absorbance of stimulated cells} - \text{background values}}{\text{total enzyme content} - \text{background values}}$$

Comments

1. Antibodies to all components of T cell receptor complex can be used.
2. It is necessary to perform a pilot experiment to establish the optimal dilution of each batch of anti-T cell receptor complex antibody.
3. Phenylmethanesulfonyl fluoride is highly toxic and proper handling is necessary.
4. Perform all assays in triplicate.

XII. Activation by phorbol myristate acetate and ionophore[21]

Materials and reagents

- Cytotoxic T lymphocytes (CTL)
- PBS
- RPMI 1640 medium supplemented with 10% FCS and antibiotics
- 1% (v/v) Triton X-100
- 1 mg/ml calcium ionophore A23187 (Sigma)
- 1 mg/ml phorbol myristate acetate (PMA; Sigma) in dimethylsulfoxide (Sigma)
- 20 m*M* *N*-α-benzyloxycarbonyl-L-lysine thiobenzyl ester (Calbiochem)
- 22 m*M* 5,5′-dithio-*bis*(2-nitrobenzoic acid) (Pierce Chemicals)
- 0.1 *M* phenylmethanesulfonyl fluoride (Sigma) in dimethylsulfoxide (Sigma)
- 96-well round-bottom microtiter plates (Dynatech)
- 37°C humidified CO_2 incubator
- Water bath
- 12 × 75-mm polypropylene culture tubes
- Centrifuge
- 50-ml conical centrifuge tubes
- Ice bath
- Spectrophotometer

Protocol

1. Dilute 1 mg/ml PMA to 40 ng/ml in complete RPMI 1640 medium immediately before use.
2. Dilute 1 mg/ml calcium ionophore to 2 μg/ml in complete RPMI 1640 medium immediately before use.
3. Add 50 μl of CTL (2×10^6/ml) in complete RPMI 1640 medium to each well. Add another 50 μl of complete RPMI 1640 medium to each well,

except those used for the determination of total granule release. Add 40 µl of complete RPMI 1640 medium and 10 µl of 1% Triton X-100 into those wells.

4. To all test wells add 25 µl of 40 ng/ml PMA and 25 µl 2 µg/ml calcium ionophore.
5. Incubate plates in a 37°C CO_2 incubator for 4 h.
6. Centrifuge the plates at 200 × *g* for 5 min at 4°C.
7. Transfer 50 µl of supernatant to 12 × 75-mm polypropylene culture tubes.
8. Prepare substrate solution:
   500 µl 20 m*M* *N*-α-benzyloxycarbonyl-L-lysine thiobenzyl ester
   500 µl 22 m*M* 5,5′-dithio-*bis*(2-nitrobenzoic acid)
   500 µl 1% Triton X-100
   48.5 ml PBS
9. Add 950 µl of substrate solution to supernatant aliquots and incubate at 37°C in water bath.
10. Move plates to an ice bath.
11. Immediately add 1 µl of 0.1 *M* phenylmethanesulfonyl fluoride to each tube.
12. Add 1 ml of PBS to each tube.
13. Measure the absorbance of samples at 412 nm.
14. Calculate percentage values by the same formula as in previous technique.

Comment

Both PMA and calcium ionophore are toxic and mutagenic; proper handling is necessary.

References

1. Fernandez-Botran, R. and Větvička, V., *Reagents in Cellular Immunology,* CRC Press, Boca Raton, 1995.
2. Henkart, P. A. and Yue, C. C., The role of cytoplasmic granules in lymphocyte cytotoxicity, *Prog. Allergy,* 40, 82, 1988.
3. Henkart, M. P. and Henkart, P. A., Lymphocyte mediated cytolysin as a secretory phenomenon, *Adv. Exp. Med. Biol.,* 146, 227, 1982.
4. Pasternack, M. S., Verret, C. R., Liu, M. A., and Eisen, H. N., Serine esterase in cytolytic T lymphocytes, *Nature,* 322, 740, 1986.
5. Simon, M. M., Prester, M., Kramer, M. D., and Fruth, D., An inhibitor specific for the mouse T cell-associated serine proteinase 1 (TSP-1) inhibits the cytolytic potential of cytoplasmic granules but not in intact cytolytic T cells, *J. Cell. Biochem.,* 40, 1, 1989.
6. Munger W. E., Berrebi, G. A., and Hankart, P. A., Possible involvement of CTL granule proteases in target cell DNA breakdown, *Immunol. Rev.,* 103, 99, 1988.
7. Russell, J. H., Internal disintegration model of cytotoxic lymphocyte-induced target damage, *Immunol. Rev.,* 72, 97, 1983.

8. Miller, R. A., Quantitation of functional T cells by limiting dilution, in *Current Protocols in Immunology,* Coligan, J. E., Kruisbeek, A. M., Margulies, D. H., Shevach, E. M., and Strober, W., Eds., John Wiley & Sons, New York, 1998, 3.15.1.
9. Bromelow, K. V., Galea-Lauri, J., O'Brien, M. E., and Souberbielle, B. E., A highly sensitive whole blood natural killer cell assay, *J. Immunol. Meth.,* 217, 177, 1998.
10. Martz, E., DNA fragmentation and cytolysin assayed by ^{3}H-thymidine, in *Cytotoxic Cells. Recognition, Effector Function, Generation, and Reagents,* Sitkovsky, M. V. and Henkart, P. A., Eds., Birkhauser, Boston, 1993, 468.
11. Howell, D. M. and Martz, E., Nuclear disintegration induced by cytotoxic T lymphocytes, *J. Immunol.,* 140, 689, 1988.
12. Willie, A. H., A new classification separating apoptosis from necrosis, in *Cell Death in Biology and Pathology,* Bowen, I. D. and Lockshin, R. A., Eds., Chapman & Hall, London, 1981, 9.
13. Kerr, J. F. R., Willie, A. H., and Currie, A. R., Apoptosis: a basic biological phenomenon with wide-ranging implications in tissue kinetics, *Br. J. Cancer,* 26, 239, 1972.
14. Willie, A. H., Glucocorticoid-induced thymocyte apoptosis is associated with endogenous endonuclease activation, *Nature,* 284, 555, 1980.
15. Willie, A. H., Cell death, *Int. Rev. Cytol. Suppl.,* 17, 755, 1987.
16. Duke, R. C., Cohen, J. J., Boehme, S. A., Lenardo, M. J., Surh, C. D., Kishimoto, H., and Sprent, J., Morphological, biochemical, and flow cytometric assays of apoptosis, in *Current Protocols in Immunology,* Coligan, J. E., Kruisbeek, A. M., Margulies, D. H., Shevach, E. M., and Strober, W., Eds., John Wiley & Sons, New York, 1998, 3.17.1.
17. Abrams, S. I. and Russell, J. H., Target cell detachment assay, in *Cytotoxic Cells. Recognition, Effector Function, Generation, and Reagents,* Sitkovsky, M. V. and Henkart, P. A., Eds., Birkhauser, Boston, 1993, 478.
18. Henkart, P., Measurement of cytolysin hemolytic activity, in *Cytotoxic Cells. Recognition, Effector Function, Generation, and Reagents,* Sitkovsky, M. V. and Henkart, P. A., Eds., Birkhauser, Boston, 1993, 484.
19. Yue, C. C., Reynolds, C. W., and Henkart, P. A., Inhibition of cytolysin activity in granules of large granular lymphocytes by lipids: evidence for a membrane insertion mechanisms of lysis, *Mol. Immunol.,* 24, 647, 1987.
20. Henkart, P. A., Millard, P. J., Reynolds, C. G., and Henkart, M. P., Cytolytic activity of purified cytoplasmic granules from cytotoxic rat large granular lymphocyte tumors, *J. Exp. Med.,* 160, 75, 1984.
21. Taffts, R. and Sitkovsky, M., Granule enzyme exocytosis assay for cytotoxic T lymphocyte activation, in *Current Protocols in Immunology,* Coligan, J. E., Kruisbeek, A. M., Margulies, D. H., Shevach, E. M., and Strober, W., Eds., John Wiley & Sons, New York, 1998, 3.16.1.
22. Pasternack, M. S. and Eisen, H. N., A novel serine esterase expressed by cytotoxic T lymphocytes, *Nature,* 314, 743, 1981.

chapter eleven

Mixed lymphocyte reaction[1]

Contents

Mixed lymphocyte reaction (MLR) is a proliferative response of two allo 1geneic T cell populations to being mixed together.[2] The major stimuli in this type of reaction are MHC antigens present on the surface membrane of interacting T lymphocytes. This is the most commonly used technique for investigating the immunological response of lymphocytes. There are three main characteristics of the MLR: synthesis of macromolecules, blast transformation, and proliferation. Cells responding in MLR are mostly $CD4^+$ lymphocytes. Since they preferentially recognize MHC class II molecules, class II antigens are the primary stimulators. In mice, we can also demonstrate an MLR against the *Mls* locus. Besides determination of the general level of immune responsiveness, this reaction is also used for determination of the MHC compatibility.

Two different types of MLR are recognized — unidirectional (also called one-way reaction) and bidirectional. In unidirectional MLR, one cell population serves as antigen (so-called simulator cells) and is used only after inhibition of its own proliferation. Therefore, only the proliferative response of the second cell population (so-called responder cells) is determined. In bidirectional MLR, proliferative response of both cell populations is measured, as neither cell population is inhibited from dividing.[3]

I. Unidirectional MLR

Materials and reagents

- Responder cell population
- Stimulator cell population, treated with mitomycin C (see Chapter 1)

- RMPI 1640 medium supplemented with 10% FCS and antibiotics
- 96-well tissue culture plate
- CO_2 incubator (humidified, set at 37°C and 5% CO_2)

Protocol

(*Note:* All materials and reagents must be sterile and proper aseptic technique must be used when handling the cells.)

1. Prepare simulator cell population treated with mitomycin C, as described in Chapter 1, and dilute cells in medium.
2. Prepare responder cell population and dilute cells in medium.
3. To a 96-well plate add 0.5, 1 and 2×10^5 responder cells per well in 0.1 ml of medium. Use at least 3 wells per concentration.
4. Add 5×10^6 simulator cells/well in 0.1 ml of medium.
5. Incubate in a humidified 37°C, 5% CO_2 incubator for 4 to 6 days.
6. Evaluate the proliferation either by MTT assay or by incorporation of [^{3}H] thymidine (see Chapter 1).

Comments

1. An entire cell population of splenic, thymic, or lymph nodes, or isolated T lymphocytes can be used as both effector and simulator cells. However, it is better to use isolated and purified cell populations. Similarly, any cell population with sufficient expression of MHC class II antigens can serve as stimulator cells. Therefore, dendritic cells and macrophages are good MLR stimulators.
2. These two cell populations must differ in MHC class II or *Mls* loci.
3. Stimulator cell population can also be irradiated instead of treated with mitomycin C. However, we found that irradiation resulted in less consistent results, as the sensitivity to ionizing irradiation varied widely. In addition, an irradiation source is not always conveniently available.
4. The optimal ratio of cells should be determined for each type of responder cells. The most common concentrations are 0.5 to 5×10^5/well for responder cells and 2 to 8×10^6/well for simulator cells.[4]
5. Wells with responder and simulator cells only will serve as a negative control. The optimal background values might be obtained from wells containing a mixture of responder and mitomycin-treated syngeneic stimulator cells.
6. Individual batches of FCS differ substantially in their support of cell proliferation. The reason for these differences might be the contamination with lipopolysaccharide or other contaminants. Different lots of FCS should be screened before the experiments.
7. Proliferation can often be increased when complete Dulbecco's medium is used with 7.5% CO_2.

8. Addition of 2-mercaptoethanol has been described as beneficial by some authors.
9. Proliferation induced by mitogens might be used as a positive control.

References

1. James, S. P., Measurement of proliferative responses of cultured lymphocytes, in *Current Protocols in Immunology*, Coligam, J. E., Kruisbeek, A. M., Margulies, D. H., Shevach, E. M., and Strober, W., Eds., John Wiley & Sons, New York, 1998, 7.10.7.
2. Shalaby, M. R., Espevik, T., Rice, G. C., Ammann, A. J., Figari, I. S., Ranges, G. E., and Palladino, M. A., The involvement of human tumor necrosis factors-α and -β in the mixed lymphocyte reaction, *J. Immunol.*, 141, 499, 1988.
3. Bradley, L. M., Mixed lymphocyte responses, in *Selected Reagents in Cellular Immunology*, Mishell, B. B. and Shiigi, S. M., Eds., W. H. Freeman and Co., San Francisco, 1980, 162.
4. Strong, D. M., Ahmed, A. A., Thurman, G. B., and Sell, K. W., *In vitro* stimulation of murine spleen cells using a microculture system and a multiple automated sample harvester, *J. Immunol. Meth.*, 2, 279, 1973.

chapter twelve

Animal models of autoimmune diseases

Contents

I. Diabetes mellitus

Diabetes mellitus is a glucose metabolism disorder caused by a relative or absolute lack of insulin. Insulin deficiency results in hyperglycemia. Two principal forms of diabetes mellitus are recognized: type I diabetes and type II diabetes. Type I diabetes is an autoimmune disease that destroys the insulin-producing cells in the pancreas, whereas type II diabetes is caused by a complex biochemical disturbance of insulin receptors.

Patients suffering from type I diabetes depend on exogenous insulin intake. This disease is fatal if untreated. After several years of diabetes, numerous complications such as accelerated arteriosclerosis or diabetic retinopathy are common. A characteristic feature of this disease is the presence of autoantibodies directed toward β cells in the islets of Langerhans.[1]

II. Nonobese mice

Nonobese diabetic (NOD) mice are a good animal model for our studies of diabetes mellitus.[2] The disease begins between 5 and 8 weeks of age; by 7 months, about 50% of animals become diabetic.[3] The significant difference from the situation in humans is that in mice the disease can be transferred by T lymphocytes. Despite the fact that all NOD mouse colonies established worldwide are derivative of the original NOD mouse derived from the Jcl:ICR strain,[4] these colonies differ considerably in the incidence of diabetes. These differences are commonly attributed to environmental factors such as diet and viruses.[5] Males are generally much less prone to develop this disease.

Because of these differences in spontaneous diabetes incidence in NOD mice, chemically induced diabetes is often used as an experimental model.

III. Chemically induced diabetes[6]

Materials and reagents

- Mice
- Streptozotocin (Upjohn)
- PBS
- Syringe
- 18-G needle

Protocol

1. Prepare solution of streptozotocin in PBS (25 mg/kg) immediately before injection.
2. Inject the mice intraperitoneally (0.5 ml/mouse) with five daily injections.
3. Wait 2 to 3 weeks for development of induced diabetes.

Comments

1. Chemically induced diabetes can be induced in any mouse strain, but NOD mice are more susceptible and therefore lower doses of agents can be used.

2. Streptozotocin is rather labile in solution; therefore, it should be stored dry.
3. Alloxan[7] or cyclophosphamide (350 mg/kg) can be used instead of streptozotocin.
4. Daily dose of 25 mg/kg streptozotocin is for male mice; in female mice use 35 mg/kg.

IV. Experimental autoimmune encephalomyelitis

Experimental autoimmune encephalomyelitis can serve as a classical example of immune-mediated demyelinating disease. It is a neurological autoimmune disease produced by an immunization with myelin basic protein. The central role of T lymphocytes can be demonstrated by transfer experiments using lymph node–derived T cells[8] and by prevention of the disease by neonatal thymectomy.[9] A characteristic feature of this disease is perivascular T-helper and mononuclear cell inflammation and subsequent primary demyelination of axinal tracks in the central nervous system, resulting in progressive hind-limb paralysis.

In mice, the disease is a Th-1 cell-dependent autoimmune disease with macrophages as the effector cells. Several mouse models exist; the most common strains are SLJ, B10.PL, and Biozzi AB/H. The Biozzi AB/H strain is particularly useful, because acute as well as chronic disease can be studied on one animal model. Peptide specificity of encephalitogenic T lymphocytes is dependent on MHC class II and is summarized in Table 1.[10]

A. Preparation of spinal cord homogenate[10]

Materials and reagents

- Mice
- Squeeze bottle with 70% ethanol
- Scissors and forceps
- Cutting board or paper towels (Fisher)
- Glass tissue homogenizer
- Desiccator
- Distilled water
- Syringe
- 18-G needle

Protocol

1. Kill mice either by cervical dislocation or CO_2 inhalation. Place the mouse on a cutting board (or on paper towel) and soak it with ethanol to reduce the possibility of hair becoming airborne.
2. Decapitate the animal. Cut the vertebrate column in front of the pelvis.

Table 1 Peptide Specificity of Encephalitogenic T Lymphocytes

Strain	Protein	Peptide	Reference
SJL	MBP	89-101	11
	PLP	139-151	12
		178-191	13
	MOG	92-106	14
B10.PL	MBP	Ac-1-9	15
		35-47	16
Biozzi AB/H	PLP	56-71	17
	MOG	1-22	14
		43-47	14
		134-148	14

Source: From Smith, R. M. and Wraith, D. C. Mouse model of experimental autoimmune encephalomyelitis in *The Immunology Reagents Manual CD-ROM*, Academic Press, London, 1997, 9.

3. Expel the cord by injecting water into the vertebral canal using an 18-G needle.
4. Collect the cords into a homogenizer on ice. Add a minimal amount of water to facilitate homogenization.
5. Lyophilize the homogenate and store at 4°C in a desiccator.

B. *Induction of experimental autoimmune encephalomyelitis*

Materials and reagents

- Mouse spinal cord homogenate
- Phosphate buffer saline (PBS)
- *Mycobacterium tuberculosis* strain H37RA (Difco)
- Freund's incomplete adjuvant
- Vortex
- Syringe
- 25-G needle
- Pertussis toxin (Sigma)

Protocol

1. Prepare a mixture of mouse spinal cord homogenate (20 mg/ml in PBS) with 8 mg/ml of *Mycobacterium tuberculosis*.
2. Add an equal volume of Freund's incomplete adjuvant and vortex.
3. Inject the mice subcutaneously at the tail base, using a 25-G needle.
4. Prepare 400 ng/ml of pertussis toxin in PBS.
5. At the time of priming and 2 days later, inject the animals intraperitoneally with 0.5 ml of pertussis toxin solution.

Comments

Any of the peptides (1 to 4 mg/ml) listed in Table 1 can be used instead of mouse spinal cord homogenate. Peptides may be synthesized or purchased from commercial vendors.

C. Adoptive induction of experimental autoimmune encephalomyelitis

Adoptive transfer of *in vitro* neuroantigen-activated lymphocytes from mice immunized with encephalitogenic antigens results in induction of experimental autoimmune encephalomyelitis in SJL mice. Clinical and histological manifestations are identical to those seen in active induction.

Materials and reagents

- SJL mice
- Mishell-Dutton balanced salt solution
- PBS
- Stainless steel screen (Thomas)
- RPMI 1640 medium containing 10% heat-inactivated FCS and supplemented with 2 m*M* L-glutamine, 10 m*M* HEPES, 100 U/ml penicillin, 100 μg/ml streptomycin, 50 μ*M* 2-mercaptoethanol, and 1 m*M* sodium pyruvate
- 50-ml conical centrifuge tubes
- 25-cm tissue culture plates
- Incubator (humidified, set at 37°C and 5% CO_2)
- Syringe
- 18-G needle

Protocol

(*Note:* All materials and reagents must be sterile and proper aseptic technique must be used when handling the cells.)

1. Use SJL mice with induced experimental autoimmune encephalomyelitis.
2. About a week after immunization, sacrifice the animals and remove lymph nodes (see Chapter 1). Place the lymph nodes in Mishell-Dutton balanced salt solution.
3. Prepare the single cell suspension at a concentration of 6×10^6 cells/ml in complete RPMI medium containing 50 μg/ml of neuroantigen used for induction of disease.
4. Add the prepared cells to 25-cm tissue culture plates.
5. Transfer the plates into a CO_2 incubator and incubate at 37°C and 7.5% CO_2 for 72 h.

6. Harvest the activated lymphocytes by centrifugating the cells in 50-ml conical centrifuge tubes for 15 min at 300 × *g*.
7. Wash the pellet two times with complete RPMI medium.
8. Count the cells and determine their viability (see Chapter 1).
9. Inject 1×10^7 viable cells in 0.5 PBS intraperitoneally into untreated SJL mice.

V. Rheumatoid arthritis

Rheumatoid arthritis is the single most common autoimmune disease in man. The principal manifestations are periarticular soft tissue swelling and joint stiffness and pain. Despite significant progress in our knowledge of the disease, there is still no cure for rheumatoid arthritis. For more information about the clinical aspects of this disease, see References 18 and 29. This disease is a complex one involving an IgM and IgG autoantibody (called rheumatoid factor), with subsequent tissue damage caused by immune complex deposition. Rheumatoid factor is present in the serum of most adult patients. Rheumatoid factor is in reality an autoantibody reacting with the Fc portion of human IgG. The occurrence of the rheumatoid factor is rather confusing, as it can be found not only in rheumatoid arthritis or systemic rheumatic disease, but also in hypergammaglobulinemic patients, or even in healthy persons. Three classes of rheumatoid factor can be found: IgM, IgA, and IgG.[20] Despite the fact that the role of rheumatoid factor is not clear, the determination of the rheumatoid factor presence remains one of the main tests in clinical immunology.

No animal model demonstrates all the features of human disease. The closest spontaneous model is MRL-*lpr/lpr* mice.[21] Nevertheless, inducible experimental models such as adjuvant arthritis and collagen-induced arthritis are more commonly used. Besides these widely used models, pristine arthritis in mice,[22] cell wall fragment-induced arthritis in rats,[23] or streptococcal cell wall-induced arthritis in rats[24] have also been described.

A. Adjuvant arthritis

This experimental model was originally demonstrated more than 40 years ago,[25] more or less by accident. Pearson found an induction of arthritis in susceptible strains of rats after immunization with complete Freund's adjuvant (*Mycobacterium tuberculosis* in oil). Later findings of the T cell-mediated nature of this disease[26] caused widespread use of this experimental model. The principal histological finding is granulosa formation in joints. The lesion is transferrable to syngeneic animals by T lymphocytes from immunized rats.[27] Additional clinical features involve nodular skin lesions and iridocyclitis.

There are certain unique aspects. For example, the disease is induced by a substance not containing a defined self-antigen. Transient arthritis found in human patients receiving immunotherapy with *Mycobacterium bovis* BCG might be a human analog to the adjuvant arthritis.[28]

B. Induction of adjuvant arthritis in rats[28]

Materials and reagents

- Susceptible rats
- 10 mg/ml heat-killed *Mycobacterium tuberculosis* (strain H37Ra) suspended in incomplete Freund's adjuvant (Difco) (see Chapter 1)
- 1-ml glass syringe
- 20-G and 25-G needles
- Restraining device

Protocol

1. Prepare and shake suspension of *Mycobacterium tuberculosis* in incomplete adjutants. Draw the suspension into a 1-ml glass syringe with a 20-G needle.
2. Measure wrist and ankle joint thickness several times and calculate the average. These values will serve as a baseline at day 0.
3. Exchange the needle with a 25-G needle.
4. Immunize the rats intracutaneously with 0.1 ml of *M. tuberculosis* suspension.
5. Daily (days 10 to 24) evaluate development of arthritis by measuring wrist and ankle joint thickness.

Comments

1. Susceptible strains include Lewis, Brown Norway, WF.1N, MAXX, and Holtzman. In all these strains the incidence of adjuvant-induced arthritis should be 100%. In the Brown Norway inbred strain of rats, however, sex-related differences might be observed.
2. Commercial sources of complete Freund's adjuvant have been found to be inferior for induction of adjuvant arthritis.[29,30]
3. In some strains of rats, incomplete Freund's adjuvant, mineral oil,[31] or pristine[32] can also be used.
4. In case the success rate in susceptible strains is below 90%, change the batch of mycobacteria or the animal supplier.

C. Collagen-induced arthritis

More experimental animals (mice, rats, and primates) are susceptible to this type of experimental autoimmune disease. The clinical manifestations resulting from immunization with type II collagen include formation of collagen-specific T cells, production of autoantibodies and anticollagen antibodies, intense synovitis, and erosion of cartilage. However, the animals do not produce rheumatoid factor. Collagen-induced arthritis in the mouse was first described by J. S. Courtenay et al. (see Reference 33.) The susceptibility is strongly controlled by MHC class II molecules, particularly by I-A^q and I-A^r alleles.[34] In rats, immunodominant peptides have been localized to a particular cyanogen bromide fragment of human type II collagen.[35] Human autoantibodies bind to the same fragment of type II collagen, which makes this model a close relative to the human disease.

1. Collagen-induced arthritis in mice[36]

Materials and reagents

- Type II collagen (Sigma)
- Sterile 10 m*M* acetic acid
- Heat-killed *Mycobacterium tuberculosis* (strain H37Ra)
- Incomplete Freund's adjuvant (Difco)
- DBA/1 mice
- Magnetic stirrer
- High-speed homogenizer
- 1-ml syringe
- 26-G needle

Protocol

1. Dissolve type II collagen in 10 m*M* acetic acid (4 mg/ml) by stirring overnight at 4°C.
2. Prepare *M. tuberculosis* suspension in incomplete Freund's adjuvant at a 4mg/ml final concentration (see Chapter 1).
3. Just prior to immunization, emulsify type II collagen and adjuvant.
4. Inject mice intradermally (50 μg/mouse) at the tail base.
5. Check the development of arthritis for evidence of erythema and swelling. Repeat scoring of paws several times a week.

Comments

1. Type II collagen can also be prepared in the laboratory. This might be especially advantageous when large quantities of type II collagen are needed. This technique was originally described by Miller and Matukas[37] and is described in detail by Rosloniec et al. (see Reference 36).
2. Be sure that collagen is kept cold throughout all steps, especially during high-speed emulsification.
3. Arthritis should appear approximately 3 to 5 weeks after immunization.
4. Alternatively, arthritis development can be checked by caliper measurement of paw swelling. In this case, constant-tension calipers are used to measure thickness of an ankle.
5. The most susceptible mouse strains are DBA/1LacJ, DBA/1J, B10.RIII, B10.Q, B10.βQBr, and BUB.

2. Measurement of B lymphocyte response

Antibody titers strongly correlate with the presence of arthritis. Besides evaluation of the B cell response, T cell response to type II collagen can be also measured either by proliferation assay or by the production of IFN-γ using a solid-phase ELISA assay.[36]

Materials and reagents

- Type II collagen
- Potassium phosphate buffer
- Anti-mouse Ig antibody conjugated to an enzyme
- Substrate
- PBS containing 0.05% Tween 20 (PBS-Tween)
- 96-well U- or flat-bottom microtiter plates (Immunol; Dynatech)
- Carbonate-bicarbonate buffer, pH 9.6
- ELISA reader
- Multichannel pipette
- Plastic squirt bottle
- Paper towels
- Serum from collagen-immunized mice

Protocol

1. Dissolve type II collagen at 5 μg/ml in potassium phosphate buffer, pH 7.4, at 4°C and coat the plates.
2. For experimental design and details of the ELISA technique, see Chapter 1.

D. Systemic rheumatic disease

The etiology of this disease remains unknown. The demonstrations of autoantibodies together with reactivity to altered self-antigens strongly suggest the autoimmune character of systemic rheumatic disease. Some examples are the formation of immune complexes with subsequent complement activation (rheumatoid arthritis) or autoantibodies to DNA in systemic lupus erythematosus.

1. Agglutination

The oldest technique originally used sheep erythrocytes coated with a subagglutinating dose of rabbit antibodies.[20] Later, synthetic polymeric microspheres with absorbed polyvalent human IgG molecules were developed.[38] In general, the latex test is more sensitive and the obtained titers are generally higher.

Materials and reagents

- Latex particles (the best results are obtained with commercial kits)
- ELISA plates coated with human IgG molecules (several commercial tests are available)
- ELISA plate reader
- Patient's serum (heat inactivated)

Protocol

1. Make serial dilution of patient's serum in appropriate buffer.
2. For experimental design and details of the ELISA technique, see Chapter 1.
3. The results are expressed in IU/ml based on a standard serum dilution curve.

2. Antiperinuclear factor

Rheumatoid factor has a predictive value, but it is not completely disease specific. The introduction of antiperinuclear factor detection has greatly improved the laboratory diagnosis of rheumatoid arthritis. The antiperinuclear factor reacts with perinuclear keratohyalin granules of buccal mucosal cells.

Materials and reagents

- Buccal mucosa cells
- Patient's serum
- Sterile foam plastic sponge
- PBS
- 15-ml conical centrifuge tubes
- PT buffer (0.5% Triton-X100 in PBS)
- Mounting medium (70% glycerol in 0.05 *M* sodium barbital, with 0.5 μg/ml ethidium bromide, pH 8.6)
- Anti-human IgA, IgM, and IgG antibodies labeled with FITC
- Fluorescent microscope
- Glass slides and coverslips

Protocol

1. Remove the buccal mucosa cells from the inner cheek of the patient using sterile foam plastic sponge.
2. Bring the cells into suspension by rinsing the sponge in 10 ml of PBS.
3. Wash two times in PBS by centrifugation at 500 × *g* at 4°C for 10 min.
4. Wash once in PT buffer by centrifugation at 500 × *g* at 4°C for 10 min.
5. Repeat Step 3.
6. Resuspend the cells in PBS at 1×10^5/ml.

7. Put 15 μl of cell suspension on microscopic slide and let dry.
8. Incubate the slide with 50 μl of patient's serum (diluted 1:5 in PBS) for 90 min at room temperature in a 100% humid atmosphere.
9. Wash three times in PBS.
10. Incubate for 30 min with FITC-labeled anti-human IgA, IgM, and IgG antibody.
11. Repeat Step 9.
12. Mount the slides in mounting medium.
13. Evaluate under the fluorescent microscope.

Comments

1. Ethidium bromide is a potential carcinogen. Use all precautions.
2. The publication "An International Reference Preparation of Rheumatoid Arthritis Serum" (WHO branch of the CLB, P.O. Box 9190, 1006 AD Amsterdam, the Netherlands) is recommended as the standard.
3. Positive and negative serum must always be used.

3. Reactive arthritis

Reactive arthritis belongs to the spondylarthropaties. Primary bacterial infection is one of the decisive factors for development of reactive arthritis. As the differential diagnosis between the reactive arthritis and other joint diseases is often difficult, a demonstration of infection by serology clearly corroborates the diagnosis. The most common forms of reactive arthritis are uroarthritis triggered by *Chlamydia trachomatis* and enteroarthritis triggered by *Campylobacter, Shigella, Yersinia,* and *Salmonella.*

Materials and reagents

- ELISA plates
- Antigen (see below)
- ELISA plate reader

Protocol

For experimental design and details of the ELISA technique, see Chapter 1.

4. Preparation of Campylobacter jejuni *antigen*

Materials and reagents

- *Campylobacter jejuni* bacteria
- *Campylobacter* blood-free selective agar plates (Oxoid)

- CCDA Selective Supplement (Oxoid)
- 0.9% NaCl
- 0.2 *M* glycine hydrochloride, pH 2.2
- Shaker
- Spectrophotometer
- Dialyzing tube
- Lyophilizator
- Sonicator

Protocol

1. Grow the bacteria on *Campylobacter* blood-free selective agar plates supplemented with CCDA Selective Supplement at 43°C for 48 h.
2. Harvest the bacteria in 0.9% NaCl and wash two times by centrifugation at 1 500 × *g* for 10 min.
3. Add 0.2 *M* glycine hydrochloride to the pellet of bacteria.
4. Shake for 30 min at room temperature.
5. Centrifuge at 8000 × *g* for 20 min at 10°C and collect supernatant.
6. Dialyze the supernatant by dialysis against water. Check the concentration and lyophilize.
7. Dissolve the lyophilized extract in PBS, sonicate and coat the plates.

5. Preparation of Yersinia *antigen*

Materials and reagents

- *Yersinia* bacteria
- Trypticase soy broth
- 0.5% formaldehyde in PBS
- PBS
- 0.1% SDS in PBS, pH 6.9
- Spectrophotometer

Protocol

1. Cultivate the bacteria in Trypticase soy broth at room temperature.
2. Harvest the bacteria by centrifugation at 3000 × *g* for 10 min.
3. Wash two times in PBS by centrifugation at 3000 × *g* for 10 min.
4. Incubate in 0.5% formaldehyde overnight at room temperature.
5. Repeat Step 3.
6. Mix 1 part of killed bacteria with 9 parts of 0.1% SDS.
7. Incubate for 1 h at 37°C.
8. Harvest by centrifugation at 1500 × *g* for 10 min.
9. Measure protein content at 280 nm using a spectrophotometer.
10. Dilute in PBS to 0.5 µg/ml.

6. *Preparation of* Borrelia burgdorferi *antigen*

Materials and reagents

- *Borrelia burgdorferi*
- BSK II medium
- PBS with 5 m*M* $MgCl_2$
- Sonicator
- Ice bath
- Spectrophotometer
- Eppendorf tubes

Protocol

1. Grow the bacteria in BSK II medium at 30°C for 5 days.
2. Harvest the bacteria by centrifugation at 10,000 × *g* for 40 min.
3. Wash five times with PBS with 5 m*M* $MgCl_2$ by centrifugation at 10,000 × *g* for 30 min.
4. Resuspend the pellet in PBS.
5. Sonicate vigorously in an ice bath.
6. Centrifuge again at 10,000 × *g* for 30 min.
7. Collect the supernatant and measure protein content at 280 nm using a spectrophotometer.
8. Aliquot in Eppendorf tubes and store at –20°C.
9. Coat the plates.

Comments

1. Optimal concentration for plate coating is 20 μg/ml.
2. Supernatant can be stored at –20°C for several months.

References

1. Vardi, P., Brik, R., and Barzilai, D., Insulin autoantibodies — reflection of disturbed self- identification and their use in the prediction of type-I diabetes, *Diabetes Metab. Rev.*, 7, 209, 1991.
2. Bach, J. F., Insulin-dependent diabetes mellitus as an autoimmune disease, *Endocrin. Rev.*, 15, 516, 1994.
3. Rossini, A. A., Modes, J. P., and Like, A. A., Immunology of insulin-dependent diabetes mellitus, *Annu. Rev. Immunol.*, 3, 289, 1985.
4. Kikutani, H. and Makino, S., The murine autoimmune diabetes model: NOD and related strains, *Adv. Immunol.*, 51, 285, 1992.
5. Leiter, E. M., Serreze, D. V., and Prochaska, M., The genetics and epidemiology of diabetes in NOD mice, *Immunol. Today*, 11, 47, 1990.
6. Kolb, H. and Kroencke, K. D., IDDM. Lessons from the low-dose streptozotocin model in mice, *Diabetes Rev.*, 1, 116, 1993.

7. Mallaise, W. J., Alloxan toxicity to the pancreatic β cell. A new hypothesis, *Biochem. Pharmacol.*, 3, 3527, 1982.
8. Paterson, P. Y., Transfer of allergic encephalomyelitis in rats by means of lymph node cells, *J. Exp. Med.*, 111, 119, 1960.
9. Arnason, B. G., Jankovich, B. D., Waksman, B. H., and Wennerston, C., Role of the thymus in immune reactions in rats. II. Suppressive effect of thymectomy at birth on reactions of delayed (cellular) hypersensitivity and the circulating small lymphocyte, *J. Exp. Med.*, 116, 177, 1962.
10. Smith, R. M. and Wraith, D. C., Mouse model of experimental autoimmune encephalomyelitis, in *The Immunology Reagents Manual CD-ROM*, Lefkovits, I., Ed., Academic Press, London, 1997, 9.
11. Sakai, K., Sinha, A. A., Mitchell, D. J., Zamvil, S. S., Rothbard, J. B., McDevitt, H. O., and Steinman, L., Involvement of distinct murine T-cell receptors in the autoimmune encepholitogenic response to nested epitopes of myelin basic protein, *Proc. Natl. Acad. Sci., U.S.A.*, 85, 8608, 1988.
12. Tuohy, V. K., Lu, Z., Laursen, R. A., and Lees, M. B., Identification of an encephalitogenic determinant of myelin proteolipid protein for SJL mice, *J. Immunol.*, 142, 1523, 1989.
13. Greer, J. M., Kuchroo, V. K., Sobel, R. A., and Lees, M. B., Identification and characterization of a second encephalitogenic determinant of myelin proteolipid protein (residues 178–191) for SJL mice, *J. Immunol.*, 149, 783, 1992.
14. Armor, S., Groome, N., Linington, C., Morris, M. M., Dornmair, K., Matthieu. J. M., and Baker, D., Identification of epitopes of myelin oligodendrocyte glycoprotein for the induction of experimental allergic encephalomyelitis in SJL and Biozzi AB/H mice, *J. Immunol.*, 153, 4349, 1994.
15. Zamvil, S. S., Mitchell, D. J., Moore, A. C., Kitamura, K., Steinman, L., and Rothbard, J. B., T cell epitope of the autoantigen myelin basic protein that induces encephalomyelitis, *Nature*, 324, 258, 1986.
16. Zamvil, S. S., Mitchell, D. J., Powell, M. B., Sakai, K., Rothbard, J. B., and Steinman, L., Multiple discrete encephalitogenic epitopes of the autoantigen myelin basic protein include a determinant for I-E class II restricted T cells, *J. Exp. Med.*, 168, 1181, 1988.
17. Amor, S., Baker, D., Groome, N. and Turk, J. L., Identification of a major encephalitogenic epitope of proteolipid protein (residues 56–70) for the induction of experimental allergic encephalomyelitis in Biozzi AB/H and nonobese-diabetic mice, *J. Immunol.*, 150, 5666, 1993.
18. Hochberg, M. C., Adult and juvenile rheumatoid arthritis: current epidemiological concepts, *Epidemiol. Rev.*, 3, 27, 1981.
19. Zvaifler, N. J., Pathogenesis of the joint disease of rheumatoid arthritis, *Am. J. Med.*, 75, 3, 1983.
20. Dorner, R. W., Alexander, R. L., and Moore, T. L., Rheumatoid factors, *Clin. Chim. Acta*, 167, 1, 1987.
21. Theofilopoulos, A. N. and Dixon, F. J., Murine models of systemic lupus erythematosus, *Adv. Immunol.*, 37, 269, 1985.
22. Thompson, S. J., Rook, G. A. W., Brealey, R. J., van der Zee, R., and Elson, C. J., Autoimmune reactions to heat-shock proteins in pristine-induced arthritis, *Eur. J. Immunol.*, 20, 2479, 1990.
23. Severijnen, A. J., van Kleef, R., Hazenberg, M. P., and van der Merwe, J. P., Chronic arthritis induced in rats by cell wall fragments of *Eubacterium* species from the human intestinal flora, *Infect. Immun.*, 5, 523, 1990.

24. Cromartie, W. J., Craddock, J. C., Schwab, J. H., Anderlen, S. K., and Jang, C. H., Arthritis in rats after systemic injection of streptococcal cells of cell-walls, *J. Exp. Med.*, 146, 1585, 1977.
25. Pearson, C. M., Development of arthritis, periarthritis, and perostitis in rats given adjuvant, *Proc. Soc. Exp. Biol.*, 91, 95, 1956.
26. Whitehouse, D. J., Whitehouse, M. W., and Pearson, C.M., Passive transfer of adjuvant-induced arthritis and allergic encephalomyelitis in rats using thoracic duct lymphocytes, *Nature*, 224, 1322, 1969.
27. Taurog, J. D., Sandberg, G. P., and Mahowald, M. L., The cellular basis of adjuvant arthritis. I. Enhancement of cell-mediated passive transfer by concanavalin A and by immunosuppressive treatment of the recipient, *Cell. Immunol.*, 75, 271, 1983.
28. Torisu, M., Miyahara, T., Shinohara, N., Ohsato, K., and Sonazaki, H., A new side effect of BCG immunotherapy - BCG-induced arthritis in man, *Cancer Immunol. Immunotherap.*, 5, 77, 1978.
29. van Eden, W., Wagenaar-Hilbers, J. P. A., and Wauben, M. H. M., Adjuvant arthritis in the rat, in *Current Protocols in Immunology*, Coligan, J. E., Kruisbeek, A. M., Margulies, D. H., Shevach, E. M., and Strober, W., Eds., John Wiley & Sons, New York, 1998, 15.4.1.
30. Best, R., Christian, R., and Lewis, D. A., Effect of particle size of dried mycobacteria on adjuvant-induced arthritis in the rat, *Agents Actions*, 14, 265, 1984.
31. Kleinau, S., Erlandsson, H., Holmdahl, R., and Klareskog, L., Adjuvant oils induce arthritis in the DA rat. I. Characterization of the disease and evidence for an immunological involvement, *J. Autoimmunol.*, 4, 871, 1991.
32. Whitehouse, M. W., Orr, K. J., Beek, W. J., and Pearson, C. M., Freund's adjuvants: relationship to arthritogenicity and adjuvanticity in rats to vehicle composition, *Immunology*, 27, 311, 1974.
33. Courtenay, J. S., Dallman, M. J., Dayan, A. D., Martin, A., and Mosedale, B., Immunization against heterologous type II collagen induces arthritis in mice, *Nature*, 283, 666, 1980.
34. Wooley, P. H., Luthra, H. S., Stuart, J. M., and David, C. S., Type II collagen induced arthritis in mice. I. Major histocompatibility complex (I region) linkage and antibody correlates, *J. Exp. Med.*, 154, 688, 1981.
35. Watson, W. C., Thompson, J. P., Terato, K., Cremer, M. A., and Kang, A. H., HLA-Dw-β gene hypervariable region homology in the biobreeding BB rat - selection of the diabetic-resistant subline as a rheumatoid arthritis research tool to characterize the immunopathologic response to human type-II collagen, *J. Exp. Med.*, 172, 1331, 1990.
36. Rosloniec, E. F., Cremer, M., Kang, A., and Myers, L. K., Collagen-induced arthritis, in *Current Protocols in Immunology*, Coligan, J. E., Kruisbeek, A. M., Margulies, D. H., Shevach, E. M., and Strober, W., Eds., John Wiley & Sons, New York, 1998, 15.5.1.
37. Miller, E. J. and Matukas, V. J., Chick cartilage collagen: a new type of alpha 1 chain not present in bone or skin of the species, *Proc. Natl., Acad. Sci. U.S.A.*, 64, 1264, 1969.
38. Cook, L. and Agnello, V., Test for detection of rheumatoid factor, in *Manual of Clinical Laboratory Immunology*, Rose, N. R., Ed., American Society for Microbiology, Washington, D.C., 1992, 762.

chapter thirteen

Buffers and media

Balanced salt solution (BSS)

The 10× BSS is made up as two stock solutions:

Stock #1:

Dextrose	10 g
KH_2PO_4	0.6 g
$Na_2HPO_4 \cdot 7\ H_2O$	3.58 g
0.5% phenol red solution	20 ml

Dissolve and bring up to 1000 ml with distilled water.

Stock #2:

$CaCl_2 \cdot 2\ H_2O$	1.86 g
KCl	4 g
NaCl	80 g
$MgCl_2$, anhydrous	1.04 g
$MgSO_4 \cdot 7\ H_2O$	2 g

Dissolve and bring up to 1000 ml with distilled water.

1. Test the 10× stocks by making a sample of 1× BSS. Mix 10 ml of stock #1 and 10 ml of stock #2, and bring up to 100 ml with distilled water. The 1× BSS should be at pH 7.2 to 7.4 and have a conductivity of 14 to 16 mS.
2. 2× and 1× BSS are obtained by appropriate dilutions of the 10×.

BSA stock solution (10%)

10 g BSA (fraction V)
44.2 ml H_2O
2 g mixed bed resin AG-501-X8(D) (Bio-Rad)
MEM (2×)

1. Add BSA to H_2O in a 100-ml beaker and let dissolve overnight.
2. Add 1 g of analytical grade mixed bed resin and incubate at 4°C for 2 h.
3. Filter the solution through paper filter.
4. Add 1 g of analytical grade mixed bed resin and repeat Steps 2 and 3.
5. Mix the BSA solution with equal volume of 2× MEM.
6. Sterilize by filtration through 0.2-µm filter, aliquot, and store at –20°C.

Carbonate-bicarbonate buffer

Na_2CO_3	1.59 g
$NaHCO_3$	2.93 g
NaN_3	0.2 g
H_2O	1000 ml

Adjust pH to 9.6. Store at 4°C for no longer than 2 weeks.

Crystal violet stain

Crystal violet	0.5 g
Ethanol	200 ml
Distilled H_2O	800 ml

1. Dissolve crystal violet in ethanol. Dilute with water.
2. Filter. Store at room temperature in dark bottle.

DEPC-treated water

Diethylpyrocarbonate (DEPC)	1 ml
Distilled H_2O	999 ml

1. In a fume hood, add 1 ml DEPC (Sigma, Cat. D-5758) to 999 ml of distilled H_2O in order to prepare a 0.1% v/v solution.
2. Let stand for at least 12 h at room temperature and then autoclave (15 min, liquid cycle).
3. Use only baked glassware. DEPC is not suitable for the treatment of Tris solutions.

Dextran solution for sedimentation of erythrocytes

Dextran T-500	30 g
Saline solution (0.85% NaCl)	For 1 l

Dissolve dextran in saline solution.

Diphenylamine reagent

Diphenylamine (Fisher)	150 mg
Glacial acetic acid	10 ml
Sulfuric acid, concentrated	150 µl
Acetaldehyde solution (see below)	50 µl

1. Place diphenylamine in a 50-ml polypropylene tube. Add glacial acetic acid and mix thoroughly.
2. Add sulfuric acid and mix again.
3. Add acetaldehyde solution and mix again.

Acetaldehyde solution

Acetaldehyde	200 µl
H_2O	10 ml

Store at 4°C for 1 year.

Ethidium bromide solution (100×)

Ethidium bromide	50 mg
Distilled H_2O	100 ml

1. Dissolve ethidium bromide in H_2O.
2. Store in a dark bottle. Dilute 1:100 for gels or staining.

Fe^{3+} cytochrome c

1. Dissolve 2.7 mg Fe^{3+} (type VI; Sigma) per ml in HBSS with Ca^{2+} and Mg^{2+}, phenol red-free.
2. Filter the solution through an 0.45-µm filter. Aliquot and store in freezer at –20°C.

Greiss reagent solutions

Solution 1	
Sulfanilamide	1 g
H_3PO_4	100 ml
Solution 2	
Naphtylethylenediamine	0.1 g
H_3PO_4	100 ml

Store both solutions in glass bottles at 4°C. Discard if not clear or if discolored.

Hanks' balanced salt solution with calcium and magnesium

$CaCl_2$	0.11 g
$MgCl_2$	95.21 mg
FCS	10 ml
Hanks' balanced salt solution	1000 ml

Store at 4°C.

McCoy's medium

McCoy's medium	85 ml
Fetal calf serum	15 ml
Sodium bicarbonate solution (7.5%)	0.6 ml
Sodium pyruvate solution (100 m*M*)	1.0 ml
Concentrated MEM essential amino acids (50×)	0.8 ml
Concentrated MEM nonessential amino acids (100×)	0.4 ml
2-Mercaptoenthanol stock (0.175 ml concentrate/50 ml H_2O)	0.1 ml
L-Glutamine (200 m*M*)	1.0 ml
L-Asparagine (10 mg/ml)	0.16 ml
L-Serine (21 mg/ml)	0.04 ml

Methylcellulose

30 g methylcellulose
500 ml H_2O
500 ml MEM (2×)

1. Slowly add methylcellulose to 300 ml of boiling water in 3-l Erlenmeyer flask. Mix for 30 min.
2. Add 200 ml of cold H_2O and then 500 ml of cold MEM (2×).
3. Shake the flask and stir for 24 h at 4°C. Aliquot and store at –20°C.

Mishell-Dutton balanced salt solution

Stock solution 1

Glucose	10 g
KH_2PO_4	0.6 g
$Na_2HPO_4{\cdot}7H_2O$	3.58 g
0.5% phenol red	10 ml
H_2O	to 1000 ml

Stock solution 2

$CaCl_2 \cdot 2H_2O$	1.86 g
KCl	4.0 g
NaCl	80.0 g
Anhydrous $MgCl_2$	1.04 g
$MgSO_4 \cdot 7H_2O$	2.0 g
H_2O	to 1000 ml

1. Prepare both stock solutions and sterilize them by filtration through 0.22 µm filter. Store at 4°C for no longer than 2 months.
2. Before experiment, prepare working solution: Add 100 ml stock solution 1 to 700 ml H_2O and then add 100 ml stock solution 2. Add H_2O to 1000 ml and use NaOH to keep pH at 7.4.

Percoll stock solution for neutrophil isolation

Percoll (Amersham Pharmacia Biotech)	90 ml
10× HBSS (Ca^{2+} and Mg^{2+} free)	10 ml

1. Mix Percoll and 10× HBSS solutions.
2. Dilute further with 1× HBSS for the preparation of density gradients. Store at 4°C.

Phosphate-buffered saline (PBS)

NaH_2PO_4	0.23 g
Na_2HPO_4	1.15 g
NaCl	9.0 g

Add H_2O to 900 ml, adjust to pH 7.2 to 7.4 using 1 *M* HCl or 1*M* NaOH. Add H_2O to 1000 ml.

PBS gel buffer

EDTA	33.62 mg
Gelatin (Sigma Type 1)	100 mg
Glucose	90 mg
PBS 10× concentrated	10 ml

Add distilled H_2O to 100 ml and adjust pH to 7.4.

Potassium phosphate buffer

K_2HPO_4	22.62 g
KH_2PO_4	0.49 g
H_2O	1000 ml

Adjust pH to 7.4. Use 0.1 *M* KH_2PO_4 if pH is too high, or 0.1 *M* K_2HPO_4 if pH is too low. Store at 4°C for no longer than 2 months.

0.1 M Sodium acetate buffer

Stock A:	
Sodium acetate	51.7 g
H_2O	750 ml
Stock B:	
Glacial acetic acid	21.7 ml
H_2O	728.3 ml
Stock C:	
3 *M* NaCl	1000 ml

1. Lower the pH of solution A with solution B to pH 4.1.
2. Mix 200 ml of solution A with 30 ml of solution C and 770 ml of H_2O.
3. Readjust the pH with solution B to pH 5.2.

SSC (20×)

NaCl	174 g
Sodium citrate	88.2 g
NaOH, 10 *N*	To adjust pH
Distilled H_2O	1000 ml

1. Dissolve NaCl and sodium citrate in 800 ml H_2O.
2. Adjust pH to 7.0 with 10 *N* NaOH solution and adjust volume to 1 l.
3. Sterilize by autoclaving (15 min, liquid cycle).

Thioglycollate medium

Cystin	0.5 g
Agar (Difco)	0.75 g
NaCl	2.5 g
Yeast extract (Difco)	5.0 g
Casein hydrolysate	15.0 g
Thioglycolic acid	0.3 ml
0.1% Resazurin (in H_2O)	1.0 ml
H_2O	to 800 ml

1. Prepare the solution and sterilize by autoclaving.

2. Add 5.6 g of glucose dissolved in 200 ml distilled H_2O and sterilized by filtration through 0.22-μm filter.
3. Aliquot and store at 4°C.

Tris-borate buffer (TBE) for electrophoresis (5×)

Tris base	54 g
Boric acid	27.5 g
0.5 *M* EDTA (pH 8.0)	20 ml
Distilled H_2O	For 1 l

1. Dissolve Tris and boric acid in 800 ml distilled H_2O.
2. Add 20 ml of 0.5 *M* EDTA. Adjust pH to 8.0.
3. Add enough distilled H_2O to make volume 1 l.

Tris-buffered ammonium chloride

Stock A:

NH_4Cl	8.3 g
H_2O	1000 ml

Stock B:

Tris base — dissolve 20.6 g Tris base in 900 ml water; adjust pH to 7.65 with HCl. Add water to 1000 ml.

Working solution:

Mix 90 ml of Solution A with 10 ml of Solution B; adjust pH to 7.2.

Tris buffer with EDTA

Tris·Cl	1.21 g
EDTA	0.37 g
H_2O	1000 ml

Adjust pH to 7.4.

Trypan blue stain

Trypan blue	0.4 g
Distilled H_2O	100 ml

1. Dissolve trypan blue in water and filter.
2. Store at room temperature in a dark bottle. Refilter after prolonged storage.

chapter fourteen

Commercial sources

Accurate Chemical and Scientific
300 Shames Drive
Westbury, NY 11590
Tel: (800) 645-6264
Fax: (516) 997-4948
http://www.accurate-assi-leeches.com

Advanced Magnetics, Inc.
61 Mooney Street
Cambridge, MA 02138-1038
Tel: (800) 343-1346
Fax: (617) 497-6927
http://www.advancedmagnetics.com

Ambion, Inc.
2130 Woodward Street
Austin, TX 78744-1832
Tel: (800) 888-8804
Fax: (512) 651-0201
http://www.ambion.com

American Type Culture Collection
10801 University Boulevard
Manassas, VA 20110-2209
Tel: (703) 365-2700
http://www.atcc.org

Amersham Pharmacia Biotech, Inc.
800 Centennial Avenue
P.O. Box 1327
Piscataway, NJ 08855
Tel: (800) 526-3593
Fax: (877) 295-8102
http://www.apbiotech.com

Baxter Diagnostics Inc.
Scientific Products Division
1430 Waukegan Road
McGaw Park, IL 60085-6787
(708) 689-8410
http://www.baxter.com/

Becton Dickinson Immunocytometry Systems
2350 Qume Drive
San Jose, CA 95131
Tel: (800) 235-5953
Fax: (408) 954-2009
http://www.bd.com
http://www.becton.co.jp/ics/profile/menu2.htm

Boehringer-Mannheim (Roche Molecular Biochemicals)
9115 Hague Road
P.O. Box 50414
Indianapolis, IN 46250-0414
Tel: (800) 262-1640
Fax: (800) 845-7355, ext. 9409; use handset
(317) 576-2754
http://www.biochem.boehringer-mannheim.com

BioWhittaker, Inc.
8830 Biggs Ford Road
Walkersville, MD 21793
Tel: (800) 638-8174
(403)-898-7025
Fax: (301) 845-8338
http://www.biowhittaker.com

Calbiochem-Novabiochem Corporation
P.O. Box 12087
La Jolla, CA 92039-2087
Tel: (800) 854-3417
(619) 450-9600
Fax: (800) 776-0999
(619) 453-3552
http://www.calbiochem.com

Corning Science Products Division (Costar)
45 Nagog Park
Acton, MA 01720
Tel: (800) 492-7711
Fax:(978) 492-1112
http://www.corningcostar.com

Curtin Matheson Scientific Co. (CMS)
General Offices
P.O. Box 1546
Houston, TX 77251-154
Tel: (713) 820-9898
Fax: (713) 878-2444

Difco Laboratories, Inc.
P.O. Box 1058A
Detroit, MI 48640
Tel: (313) 961-0800
http://www.difco.com

Du Pont NEN Research Products
Customer Service
549 Albany Street
Boston, MA 02118
Tel: (800) 551-2121
http://www.dupont.com

Dynal Inc.
5 Delaware Drive
Lake Success, NY 11042
Tel: (800) 638-9416
Fax: (516) 326-3298

Dynatech Laboratories, Inc.
900 Slaters Lane
Alexandria, VA 22314
Tel: (800) 336-4543
Fax: (703) 548-3889

Fisher Scientific
711 Forbes Avenue
Pittsburgh, PA 15219
Tel: (800) 766-7000
Fax: (800) 926-1166
http://www.fisher1.com

Genzyme Diagnostics
One Kendall Square
Cambridge, MA 02139-1562
Tel: (617) 252-7500
Fax: (617) 252-7759
http://genzyme.shore.net

Gibco BRL (Life Technologies)
Corporate Headquarters
Life Technologies, Inc.
P.O. Box 9418
Gaithersburg, MD 20898
Tel: (800) 828-6686
Fax: (800) 331-2286
http://www.lifetech.com/

ICN Biomedicals, Inc.
3300 Hyland Avenue
Costa Mesa, CA 92626
Tel: (800) 854-0530
Fax: (800) 334-6999
http://www.icnbiomed.com/

Intergen Company
The Centre at Purchase
2 Manhattanville Road
Purchase, NY 10577
Tel: (800) 431-4505
Fax: (914) 694-1429
http://www.intergenco.com

JRH Biosciences
13804 W. 107th Street
Lenexa, KS 66215-0848
Tel: (800) 255-6032
Fax: (913) 469-5584
http://www.jrhbio.com

Kodak Chemicals
Eastman Kodak Company
25 Science Park
New Haven, CT 06511
Tel: (800) 225-5352
http://www.kodak.com

Miltenyi Biotec Inc.
1250 Oakmead Park
Suite 210
Sunnyvale, CA 94088
Tel: (800) 367-6227
Fax: (916) 888-8925
http://www.miltenyibiotec.com

Molecular Dynamics
928 E. Argues Avenue
Sunnyvale, CA 94086
Tel: (408) 773-1222
Fax: (408) 773-1493
http://www.mdyn.com

Molecular Probes, Inc.,
4849 Pitchford Avenue
Eugene, OR 97402-0469
Tel: (541) 465-8300
Fax: (541) 344-6504
http://www.probes.com

Nalge Nunc
75 Panorama Creek Drive
P.O.Box 20365
Rochester, NY 14602-0365
Tel: (800) 625-4327
Fax: (716) 586-8987
http://www.nunc.nalgenunc.com

Orpegen
Czerny-Ring 22
69115 Heidelberg
Germany
Tel: 49-6221-9105
Fax: 49-6221-9106
http://www.orpegen.com

Oxoid Inc.
800 Proctor Avenue
Ogdensburg, NY 13669
Tel: (800) 567-8378
http://www.oxoid.ca

Pharmacia and Upjohn
95 Corporate Drive
P.O.Box 6995
Bridgewater, NJ 08807
Tel: (908) 306-4400
Fax: (908) 306-4433
http://www.pharmacia.se/ns4index.html

Pharmingen (Becton Dickinson)
10975 Torreyana Road
San Diego, CA 92121-1111
Tel: (800) 848-6227 (orders)
(800) 825-5832 (technical)
Fax: (619) 812-8888
http://www.pharmingen.com

Pierce Chemical
P.O. Box 117
Rockford, IL 61103
Tel: (800) 874-3723
Fax: (815) 968-7316
http://www.piercenet.com

Polysciences, Inc.
400 Valley Road
Warrington, PA 18976
Tel: (800) 523-2575
Fax: (800) 343-3291

Promega Corporation
2800 Woods Hollow Road
Madison, WI 53711-5399
Tel: (800) 356-9526
Fax: (608) 277-2516
http://www.promega.com

R & D Systems, Inc.
614 McKinley Place N.E.
Minneapolis, MN 55413
Tel: (800) 343-7475
Fax: (612) 627-0424
http://www.rndsystems.com

Research Products International Corp.
410 N. Business Center Drive
Mount Prospect, IL 60056-2190
Tel: (708) 635-7330
Fax: (708) 635-1177

Sigma Chemical Company
P.O. Box 14508
St. Louis, MO 63178
Tel: (800) 325-3010
(314) 771-5750
Fax: (800) 325-5052
(314) 771-5757
http://www.sigma-aldrich.com

Skatron Instruments, Inc.
108 Terminal Drive
P.O. Box 530
Sterling, VA 20167
Tel: (800) 336-0191
Fax: (703) 478-5197
http://www.skatron.com

Tel-Test, Inc.
P.O. Box 1421
1511 Country Road 129
Friendswood, TX 77546
Tel: (800) 631-0600
Fax: (713) 482-1070

Thomas Scientific
99 High Hill Road at I-295
P.O. Box 99
Swedesboro, NJ 08085-0099
Tel: (800) 345-2100
(609) 467-2000
Fax: (609) 467-3087
http://www.thomassci.com

Winthrop Pharmaceuticals
Division of Sterling Drug Inc.
90 Park Avenue
New York, NY 10016
Tel: (212) 907-2000
Fax: (212) 551-7856

Index

T

U

W

Y

Z

For Product Safety Concerns and Information please contact our EU representative GPSR@taylorandfrancis.com Taylor & Francis Verlag GmbH, Kaufingerstraße 24, 80331 München, Germany